W9-CCA-633

Developing Quality Technical Information

A Handbook for Writers and Editors

Gretchen Hargis • Michelle Carey • Ann Kilty Hernandez • Polly Hughes • Deirdre Longo • Shannon Rouiller • Elizabeth Wilde

PRENTICE HALL
Professional Technical Reference
Upper Saddle River, New Jersey 07458
www.phptr.com

Editorial/Production Supervision: Wil Mara
Cover Design Director: Jerry Votta
Cover Design: Talar Boorujy
Art Director: Gail Cocker-Bogusz
Manufacturing Manager: Alexis R. Heydt-Long
IBM Consulting Editor: Susan Visser
Publisher: Jeffrey M. Pepper
Editorial Assistant: Linda Ramagnano
Marketing Manager: Robin O'Brien

Publishing as Prentice Hall Professional Technical Reference
Upper Saddle River, New Jersey 07458

Prentice Hall PTR offers excellent discounts on this book when ordered in quantity for bulk purchases or special sales. For more information, please contact: U.S. Corporate and Government Sales, 1-800-382-3419, corpsales@pearsontechgroup.com. For sales outside of the U.S., please contact: International Sales, 1-317-581-3793, international@pearsontechgroup.com.

Printed in the United States of America

First Printing

ISBN 0-13-147749-8

Pearson Education Ltd.
Pearson Education Australia Pty., Limited
Pearson Education Singapore, Pte. Ltd.
Pearson Education North Asia Ltd.
Pearson Education Canada, Ltd.
Pearson Educación de Mexico, S.A. de C.V.
Pearson Education—Japan
Pearson Education Malaysia, Pte. Ltd.

Contents

IBM Press Series—Information Management

On Demand Computing Books

On Demand Computing
Fellenstein

Grid Computing
Joseph and Fellenstein

Autonomic Computing
Murch

Business Intelligence for the Enterprise
Biere

DB2 Books

DB2 Universal Database v8.1 Certification Exam 700 Study Guide
Sanders

DB2 Universal Database v8.1 Certification Exams 701 and 706 Study Guide
Sanders

DB2 for Solaris: The Official Guide
Bauch and Wilding

DB2 Universal Database v8 for Linux, UNIX, and Windows Database Administration Certification Guide, Fifth Edition
Baklarz and Wong

Advanced DBA Certification Guide and Reference for DB2 Universal Database v8 for Linux, UNIX, and Windows
Snow and Phan

DB2 Universal Database v8 Application Development Certification Guide, Second Edition
Martineau, Sanyal, Gashyna, and Kyprianou

DB2 Version 8: The Official Guide
Zikopoulos, Baklarz, deRoos, and Melnyk

Teach Yourself DB2 Universal Database in 21 Days
Visser and Wong

DB2 UDB for OS/390 v7.1 Application Certification Guide
Lawson

DB2 SQL Procedural Language for Linux, UNIX, and Windows
Yip, Bradstock, Curtis, Gao, Janmohamed, Liu, and McArthur

DB2 Universal Database v8 Handbook for Windows, UNIX, and Linux
Gunning

Integrated Solutions with DB2
Cutlip and Medicke

DB2 Universal Database for OS/390 Version 7.1 Certification Guide
Lawson and Yevich

DB2 Universal Database v7.1 for UNIX, Linux, Windows and OS/2—Database Administration Certification Guide, Fourth Edition
Baklarz and Wong

DB2 Universal Database v7.1 Application Development Certification Guide
Sanyal, Martineau, Gashyna, and Kyprianou

DB2 UDB for OS/390: An Introduction to DB2 OS/390
Sloan and Hernandez

More Books from IBM Press

Developing Quality Technical Information: A Handbook for Writers and Editors
Hargis, Carey, Hernandez, Hughes, Longo, Rouiller, and Wilde

Enterprise Messaging Using JMS and IBM WebSphere
Yusuf

Enterprise Java Programming with IBM WebSphere, Second Edition
Brown, Craig, Hester, Stinehour, Pitt, Weitzel, Amsden, Jakab, and Berg

Welcome

Many books about technical writing tell you how to develop different parts of technical information, such as headings, lists, tables, and indexes. Instead, we organized this book to tell you how to apply quality characteristics that, in our experience, make technical information easy to use, easy to understand, and easy to find. We hope you will find our approach useful and comprehensive—and we hope that you will find the information in this book easy to use, easy to understand, and easy to find!

Is this book for you?

If you are a writer, editor, or reviewer of technical information—yes! If you write, edit, or review software information, this book might be of even more interest to you because most of the examples in it come from the domain of software. However, the quality characteristics and guidelines apply to all technical information.

Reviewers can be any of the many people who are involved in developing technical information:

- Writers
- Editors
- Visual designers
- Human factors engineers
- Product developers and testers
- Customer service personnel
- Customers (perhaps as early users)
- Managers

In general, this book assumes that you know the basics of good grammar, punctuation, and spelling as they apply to writing. It does not assume that you are familiar with what makes technical information good or bad.

How to use this book

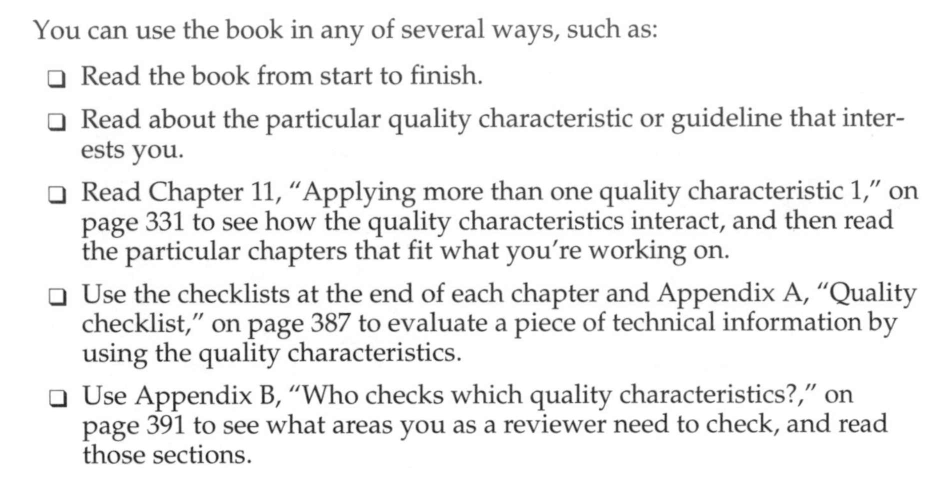

You can use the book in any of several ways, such as:

- Read the book from start to finish.
- Read about the particular quality characteristic or guideline that interests you.
- Read Chapter 11, "Applying more than one quality characteristic 1," on page 331 to see how the quality characteristics interact, and then read the particular chapters that fit what you're working on.
- Use the checklists at the end of each chapter and Appendix A, "Quality checklist," on page 387 to evaluate a piece of technical information by using the quality characteristics.
- Use Appendix B, "Who checks which quality characteristics?," on page 391 to see what areas you as a reviewer need to check, and read those sections.

Whatever your role in developing technical information, we hope that you'll use this information to build these quality characteristics into the information that you work on.

Conventions used in this book

Nine of the twelve chapters in this book deal with the quality characteristics, one per chapter. Each of these chapters has a series of guidelines about how to enhance the particular quality characteristic.

Within each guideline, this book uses examples, usually in pairs of an original passage such as you might see in technical information and a revision that demonstrates the application of the guideline. Some passages go through more than one revision. The descriptions of the guideline and of the examples aim to help you understand and implement the guideline.

In addition, each of the nine chapters ends with a checklist. This checklist indicates the items to look for when you evaluate a piece of technical information by using the guidelines for the particular quality characteristic.

Changes in this edition

The basic organization of the book and the quality characteristics remain the same. However, within each quality characteristic, we have added, reworded, deleted, or moved some guidelines and subguidelines, and we have updated many examples. For example, the following guidelines are among those that we added:

- "Organize information into discrete topics by type." (Organization chapter)
- "Facilitate navigation and search." (Retrievability chapter)
- "Ensure that all users can access the information." (Visual effectiveness chapter)

These changes resulted from several developments in technical communication:

- Greater emphasis on topic-based information and single source
- Internationalization of information and increased delivery of technical information on the Web
- The need to make technical information accessible to people with disabilities such as blindness and deafness

As with earlier developments in this field during the 20 years that these quality characteristics have been in use, the characteristics have been able to absorb the changes. This framework continues to apply to the information that we are called on to provide today.

We hope that you find this book useful in improving the quality of the information that you develop.

Gretchen Hargis
Michelle Carey
Ann Kilty Hernandez
Polly Hughes
Deirdre Longo
Shannon Rouiller
Elizabeth Wilde

Acknowledgments

The previous edition of this book (1998) came out 12 years after the edition before it (1986). Our acknowledgments for the 1998 edition mentioned over 50 people who had directly or indirectly contributed to that book in the intervening years. Now, five years later, our list is much shorter, but we would still like to acknowledge the contributors to this book's predecessor, *Producing Quality Technical Information*: the lead writer Morris Dean and the writers, editors, and designers who worked with him—Dewey Beaudette, Fred Bethke, the late Bill Calhoun, Polly Hughes, John Hurd, Terese Johnson, Kacy Keene, Lori Neumann, and Linda Stout. *Producing Quality Technical Information* was also recognized as a classic in technical communication by the *Journal of Computer Documentation* (August 2002).

For this edition, we thought that the work would go quickly. After all, we had only to bring in changes since 1998. Nevertheless, life itself was much less accommodating for this revision. Many of the writers experienced major joys and sorrows, which take priority over even a revision schedule for a book: a marriage, two births, two deaths, a divorce, and life-threatening illnesses, not to mention the increased pressures of work.

Not all of the writers who started this race finished it, and some writers ran the race as a relay, handing off work while on a leave of absence or otherwise occupied. We thank Jim Ramaker for his heavy lifting to reorient a chapter, and we regret that he could not finish. We thank Moira Lanyi for her extraordinary versatility in helping to revise a chapter, review the book, create artwork, format the book, bring order to our production library, and review the final document. We thank Kristi Ramey for her efficient handling of updates to two chapters, her helpful tips for using FrameMaker, and her eagerness to help. We thank Bill Bischoff for stepping in to format the book, with lightning speed, to expedite a review. We thank Colleen Enright for her immense FrameMaker competence, which she generously applied to helping with the index. We thank Bob Rumsby, who rescued us from formatting problems at the eleventh hour, and Gary Faircloth, who also helped greatly with final formatting.

Reviewers of the published book and of drafts for this edition gave us many useful comments. Our thanks to Ann Jackson for her ideas for icons and examples, Debbie Mayhew for her help in revising the information about usability testing, Teresa Stoll and Deborah Slayton for their contributions to the information about visual effectiveness, Patrick Kelley for reviewing an early draft, Jacqueline Kushner for sharing her experiences from using the book in a college class, Jon Swedlow for his suggestions for improvements, Thomas Sharp for his advice and counsel, Eric Radzinski for reviewing the final document, and readers who wrote their comments on amazon.com.

Thanks to Lindsay Bennion, in spring 2002 we had the rare opportunity to work with Indiana University graduate students in a class called "Effective Writing in Instructional Technology," which used the 1998 edition of *Developing Quality Technical Information* as the text. We especially thank Professor Elizabeth Boling, assistant instructor Chuck Stewart, and students Jonathan Bartling, Lori Ceremuga, Rebecca Ellis, Mark Kenefick, and Jim Steele for the insights that we gained into the real application of our book in a class situation and for letting us participate in a couple of their projects.

Lori Fisher supported this endeavor, coming up with solutions at critical times. Shawn Benham helped sustain us when our energy was running low. Frank Eldredge perservered in helping us work through difficult publishing issues. Barbara Isa and Tom Vogel helped us get on the right road with the contract, and Susan Visser and Terry McElroy helped with contract issues.

Lastly, we thank our families and loved ones for their understanding when we worked seemingly endless evenings and weekends on this revision.

Again, if we missed or slighted someone in these acknowledgments, we didn't mean to.

Gretchen Hargis
Michelle Carey
Ann Kilty Hernandez
Polly Hughes
Deirdre Longo
Shannon Rouiller
Elizabeth Wilde

Chapter 1

Quality technical information

What is technical information? And how is quality technical information different from just technical information?

Technical information is information about a technical subject, usually for a particular audience and for a stated purpose. When such information accompanies a product, it tends to be practical but usually not for the purpose of selling the product. Technical information can also explain or describe something in a science (whether computer science, physical science, or social science), trade, or profession.

In terms of the traditional forms of discourse, technical information involves primarily exposition and description, but it can also involve narration and even persuasion. Scenarios help elucidate technical information, and task information is more effective if it appeals to users' motivations for doing the tasks.

Technical information is not leisure reading for most people. Rather, they turn to technical information out of a need to know something, to solve a problem, probably as part of their job. Often they have little time to find the information that they need.

Unfortunately, much technical information has a reputation as the last thing people want to read or the last thing to turn to for help. Technical information has become almost laughable as boring, incomprehensible, unusable, and impossible to find.

Quality technical information has characteristics that deal with these problems. Writers of technical information need to make the information easy to use, easy to understand, and easy to find.

Content that is structured as separate types of information (specifically, task, concept, and reference topics) can help move technical information toward these characteristics and also facilitate the reuse of information.

This chapter provides an overview of the following subjects:

- Quality technical information
- The process of developing quality technical information
- Topic-based information as a way of writing and reusing quality technical information

What is quality technical information?

Quality has been a major concern of businesses for many years, particularly in terms of the characteristics of the output. Products compete not only on how much they cost and whether they're available when and where needed, but also on the quality of what they offer. This competition crosses national boundaries when markets span the globe. For example, in the 1970s car manufacturers in the United States needed to reconsider the level of quality in their cars when car manufacturers from Japan began to sell what people perceived as higher-quality cars, cars that offered features such as better fuel economy and more reliability.

Quality characteristics for technical information must also reflect what users expect and want from the information. Based on comments from users and on experience in writing and editing technical information, the authors of this book have found that quality technical information has these characteristics:

Easy to use	
Task orientation	A focus on helping users do tasks that are associated with a product or tool in relation to their jobs.
Accuracy	Freedom from mistake or error; adherence to fact or truth.
Completeness	The inclusion of all necessary parts—and only those parts.
Easy to understand	
Clarity	Freedom from ambiguity or obscurity; the presentation of information in such a way that users understand it the first time.
Concreteness	The inclusion of appropriate examples, scenarios, similes, analogies, specific language, and graphics.
Style	Correctness and appropriateness of writing conventions and of words and phrases.
Easy to find	
Organization	A coherent arrangement of parts that makes sense to the user.
Retrievability	The presentation of information in a way that enables users to find specific items quickly and easily.
Visual effectiveness	Attractiveness and enhanced meaning of information through the use of layout, illustrations, color, typography, icons, and other graphical devices.

From the users' point of view, each of these characteristics is necessary to describe quality technical information. Quality is a composite of several characteristics.

Relationships among the quality characteristics

Ideally, each characteristic should be discrete and not overlap any other characteristic. With such separateness, writers could easily agree on where each guideline and quality problem fits in the spectrum of quality characteristics. Without such separateness, conventions about how to deal with overlapping areas are needed.

One convention that the authors of this book established was to group these characteristics by their ability to make information easy to use, easy to understand, and easy to find. Each group focuses on the primary area that a quality characteristic affects. However, a characteristic might also influence another group. For example, accuracy contributes primarily to making technical information easy to use, but it can also help make the information easy to understand.

These groups help to explain the quality characteristics and make them easier to remember. They relate the characteristics to the basic tasks that users need to do with technical information: find it, understand it, and use it. Success with each task, without having to spend much time or effort, is essential to a user's feeling of satisfaction with the information.

The easy-to-use group of quality characteristics deals with the actual application of the information. That application might entail, for example, installing a product, adding a row to a table, or writing a line of code. To be able to apply the information, the user must have found and understood the information. Unlike usability, *use* in "easy to use" means *apply*, not just any kind of contact with the information. Usability, like quality, is an umbrella concept.

Order of the groups

Although the user needs to first find the information, this book describes the easy-to-use group first for two reasons:

- Using the information, whether to learn or to do something, is the primary reason why a user tries to find and understand information. Finding the information or understanding it is usually not the user's main purpose.

- The primary audience for this book is writers and editors of technical information rather than the users of technical information. Writers and editors need to give foremost importance and consideration to making the information easy to use.

So rather than reflecting the sequence of user tasks for technical information, the order reflects the general importance of the quality characteristics to users.

This book discusses how to achieve quality by applying each of the three quality characteristics in each of the three groups as you develop technical information. Chapter 11, "Applying more than one quality characteristic," on page 331 treats examples of using the quality characteristics together rather than just one at a time.

Quality characteristics compared with elements and guidelines

Quality is difficult to identify, codify, or imitate. You cannot simply use headings, lists, and tables in the same way as they appear in a document that you admire and achieve the same degree of quality. Yet, when you apply wise *guidelines* to the use of such *elements*, you can improve quality. In this book we use these terms with these meanings:

- *Guidelines* are brief directives about what to do or not do to achieve the characteristics of quality. These guidelines apply to the elements.
- *Elements* are the units that physically constitute technical information. An element can be as small as a word and as large as a tutorial.

Paragraphs evolved as the primary element for conveying meaning in writing. Works of fiction tend to use paragraphs almost exclusively; they might also use chapter titles and a table of contents, as clues to their structure. Technical information, however, requires more kinds of elements and liberal use of them. The paragraph becomes much less important as an element for conveying information. It becomes shorter and is broken up with lists, tables, and other means of differentiating information.

The quality characteristics of technical information are different from elements in these ways:

- Each quality characteristic expresses an innate goodness. The characteristics are not neutral, as are the elements.

Lists, for example, do not in themselves constitute quality technical information. Writers can use lists to achieve certain quality characteristics, such as clarity and retrievability. However, writers can also misuse lists—hence the need for guidelines about what to do and not do with lists.

- The information does not consist of the quality characteristics.

  Writers cannot, for example, turn clarity on and off with a markup tag or a word-processor option. Elements, however, do have recognizable limits and often have corresponding markup tags or word-processor options.

The elements span the quality characteristics (and vice versa), as shown in Appendix C, "Quality characteristics and elements," on page 397.

Other possible quality characteristics of technical information

The quality characteristics that this book covers are not the only possible quality characteristics of technical information. You might think of some other terms to define quality technical information, such as:

Conciseness Brevity, succinctness; saying a lot in a few words.

Consistency Using the same elements or content where appropriate; agreement or logical coherence among parts.

Preciseness Clear expression; correctness to a fine degree.

Readability Ease of reading words and sentences.

Relevance Appropriateness to a subject.

Simplicity Freedom from complexity.

However, clarity includes these characteristics. Clarity is broad enough to cover many characteristics and yet narrow enough to be different from the other eight characteristics in this book.

Some of these possible quality characteristics are part of more than just clarity. Consistency and simplicity, for example, are also part of style, organization, and visual effectiveness.

Some other possible characteristics are:

Adequacy Just enough information; the right amount and kind of information to meet a need.

Correctness Freedom from mistakes or errors; this term is often used in technical writing in regard to matters of style.

Honesty Truthfulness, freedom from fraud or deceit.

Usefulness Capability of being used to advantage or of being of service.

Again, the characteristics that we have chosen can cover these:

- Accuracy includes correctness and honesty.
- Completeness includes adequacy.
- Style includes correctness.
- Task orientation includes usefulness.

The nine quality characteristics in this book are a complete set for developing quality technical information.

Using the quality characteristics to develop quality technical information

Quality is sometimes thought to be incompatible with schedule and costs. Writers think they don't have enough time or a big enough budget to develop quality technical information. After all, quality is not like a coat of paint that can be added at the end of a development process, making the product look good but possibly covering up deeper problems. Various standards (such as ISO 9001) have been created to help ensure that the development process leads to a quality product.

Developing quality technical information requires many skills, such as researching, interviewing, designing, prototyping, writing, editing, reviewing, and usability testing. To cover these skills, probably more than one person is needed, though a writer can have many of these skills. This book concentrates on the writing, editing, testing, and reviewing skills.

Writing quality technical information is a complex task. You probably cannot achieve quality in one quick rush of words. In fact, a very important part of the writing process isn't writing at all, but preparing to write.

Although any of the quality characteristics can be considered at any time, focusing on all of them equally at all times would be overwhelming. Figure 1 shows the general development process and the quality characteristics that are probably most appropriate along the way.

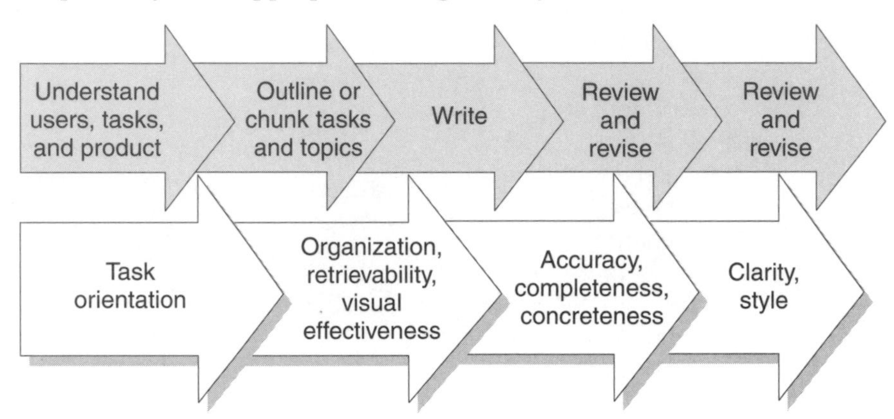

Figure 1. Some quality characteristics require more consideration than others at particular points during the development cycle.

Preparing to write: understanding users, tasks, and the product

As a writer, you need to understand not only the product, writing tools, and the writing craft, but also certain characteristics of the typical user:

- ❑ Education and training
- ❑ Knowledge of the subject or related subjects
- ❑ Work environment

You need to build on the user's knowledge and speak the user's language, while considering also the needs of the disabled (people who have limited vision, hearing, mobility, or attention) to use the information. You can do an *audience analysis* to better understand your users and a *task analysis* to better understand their tasks. You can monitor users as they work, interview them about their tasks, or survey a set of users. In a task analysis, you can focus on high-level tasks as well as on more detailed tasks.

The results of a task analysis can help you understand:

- ❑ Which tasks are most important to users
- ❑ Which tasks they spend the most time on
- ❑ Which tasks are most tedious or most trivial
- ❑ What type of organization is appropriate
- ❑ How much detail to include
- ❑ What kinds of examples and visual elements to use

The more that you understand about the users and their tasks, the easier it is for you to write relevant, task-oriented information.

If what you are writing about is new to you, pay close attention to your own questions and areas of confusion as you learn about the subject. The information that you write might need to serve both the novice and the experienced user. You can use your own experience to help you determine what information new users need and how best to present it.

You need to understand how the product fits into what the user does and what new tasks the product requires of users. Above all, the users of your information are probably less interested in using the information or the product that it represents than in doing their own work.

Many of the guidelines in this book speak of users as the reference point for decisions about what to write and how to write it.

Writing and rewriting

As with other kinds of writing, quality technical writing is the result of a process of refinement. When you are first writing about a subject, just getting your thoughts on the screen or paper might be enough of a challenge. Striving at that point for quality characteristics such as retrievability and style might be more inhibiting than helpful.

You need not concentrate on achieving all of the quality characteristics at once. Some are more appropriate than others, depending on your approach and where you are in the development cycle. If you are using a top-down approach (going from an outline to filling it in), you probably need to keep in mind the guidelines for task orientation and organization. If you are using a bottom-up approach (writing pieces of information and then fitting them together), the guidelines for clarity and visual effectiveness might be appropriate.

Eventually, as you develop the information, the guidelines for all the quality characteristics should come into play. Many of them need frequent consideration. Consider the clarity guideline "Use technical terms only if they are necessary and appropriate" on page 141. Early in your writing, you might decide to use certain technical terms and then reconsider their use as you get feedback from users and other reviewers.

You need to revise sentences, paragraphs, procedures, and larger blocks of information, always bringing them closer to the goal of being task oriented, accurate, complete, clear, concrete, stylistically correct and appropriate, well organized, easily retrievable, and visually effective.

Reviewing, testing, and evaluating technical information

You can evaluate your own information by applying the guidelines for the quality characteristics. At the end of each of the nine chapters about the quality characteristics, you can see in the checklist the problems to look for concerning each guideline. You can also have another person use the checklists to evaluate what you have written, or you can use the checklists when you evaluate what someone else has written.

These individual checklists are gathered into one checklist in Appendix A, "Quality checklist," on page 387. You can put your findings from the individual checklists into this summary checklist, to get a total picture.

All of the checklists provide columns where you can give a numerical evaluation (from 1 to 5) based on the number and kinds of problems that you find. These numerical evaluations can help you see the areas that most need revising.

The major revision cycles in the process of developing technical information probably result from reviews by other people who are involved in the project, such as developers, designers, editors, human factors engineers, users, and marketing people. You can use the checklist for the quality characteristics in Appendix B, "Who checks which quality characteristics?," on page 391 to help ensure that the different types of reviewers evaluate the appropriate quality characteristics.

Chapter 12, "Reviewing, testing, and evaluating technical information," on page 357 deals with this subject in more detail.

Writing task, concept, and reference topics

Writers sometimes duplicate work to supply the same or similar information in different media or in different collections of the information (such as online help and a printed document). Using the same source (single source), with few or no changes, for different outputs can save time and also help ensure that the information is accurate and consistent wherever it appears. Single source also helps reduce the time to review the information, translate it, and maintain it.

Establishing a unit of reuse

How can you develop information once and then get maximum reuse from it? One way is to make the unit of reuse small enough to work in several places and yet large enough to have meaning. One sentence, for example, is probably too small to manage as a reusable element, and one chapter is probably too large to work in more than one place. The unit that is probably most useful is a *topic* (sometimes also called a *module*, a *chunk*, or an *article*), which reflects a particular content and is neutral with respect to where or how it is presented. Basically, a topic consists of a title and some content.

A topic can be refined as one of three types, as shown in Figure 2:

Task Step-by-step instructions on how to do an action, plus the rationale or context for the action.

Concept Background information that users need to know to do the tasks, including extended definitions and explanations.

Reference Facts that support a task, such as programming syntax and restrictions.

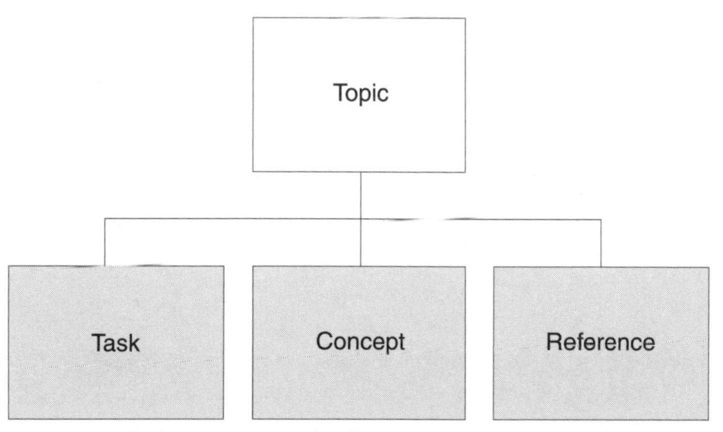

Figure 2. Task, concept, and reference are three types of topics.

The particular content for each topic type is unique to that type. A concept, for example, should include a definition and might add one or more of the following elements:

- Graphic
- Analogy
- Example
- Scenario

For most collections or sequences of topics, tasks are the main type of topic. They are probably the most important type of topic for users, but they need the support of concepts and reference information. Novice users are likely to need concepts, and experienced users are likely to need reference information.

Contextual information can be considered a separate type of topic or a combination of the main three. Similarly, tutorials are primarily task topics plus appropriate concepts.

Links between the topics can establish a network that suits the particular collection of topics and its users. Metadata about the topics can also enable dynamic selection of topics by users. Such metadata might include characteristics such as:

- Product and release
- Task relationships such as parent task and child task
- User role and skill level
- Operating system

Metadata about the progress of a topic (such as review and approval status) through a workflow can help with tracking the topic.

This book encourages approaching technical information as topics, in the interest of making quality technical information easier to achieve.

Restructuring technical information

Technical writers sometimes must deal with existing information that was written by someone else, perhaps many years ago. As in other professions, particularly those in the computer industry, technical communication has undergone many changes. In the 1970s and early 1980s technical writers wrote mainly books. When they started writing online information, they usually wrote this information separately from the books.

In recent years the emphasis has been on some form of online information, whether contextual information or concept, task, or reference topics in an information center. Even technical information in the form of books has become linear documents that are presented in an online format rather than printed. Also, the move to XML as a source language encourages the use of single source across different types of output.

Restructuring information is a special type of revision. Typically, the old information has these characteristics:

- The smallest manageable unit of information is a chapter.
- Each chunk mixes conceptual, task, and reference information.
- Any place in the information can refer to any other place inside or outside the information. The information is intertwined like spaghetti.
- The writing assumes a certain order and output, as in phrases such as "The following chapter describes" and "as in the previous section."

However, if the information is to be used in various types of output, the information must be restructured to give it the following characteristics:

- The smallest manageable unit of information is a topic, with a title, which usually has no internal headings and can be at any level of any hierarchy.
- Each topic contains only one type of information, such as a concept, task, or reference.
- The references to other topics are limited to certain places in a topic.
- The wording is neutral in terms of the type and order of the output.

Therefore, writers need to restructure, or rearchitect, information so that it can integrate with new information that is being written. This book can also help with this process of restructuring existing information.

Part 1

Easy to use

Ease of use is very broad. Some people equate the term with usability, which they consider as broad as quality itself. Here we limit ease of use to the primary characteristics that determine whether users can actually apply the information. We distinguish this ability from being able to find and understand the information.

Chapter 2

Task orientation

Don't tell me how it works, tell me how to use it. —A customer

Task-oriented writing is writing in terms of how the user does the task. You rarely help your users when you tell them only how a product works or how it is structured internally. Your users have a job to do, so they need practical information—how-to information.

You need to understand the tasks that you're writing about from your users' perspective. Do a task analysis to determine which tasks are most important to each group of users, which tasks are most frequent, and which tasks are most difficult. Make a list of the high-level tasks that users will do with your product.

You can divide high-level tasks, such as getting started with the product, into groups of lower-level tasks, such as installing the product and setting up the product. Each of these tasks might also be divided into still lower-level tasks until you have groups of manageable tasks that make sense to the user. For example, opening a bank account is a discrete task that makes sense for a user to do. However, the action of typing an address might be a step of a task, but it is not a task on its own.

Task topics, whether high-level or low-level, are the most important types of topics for users because tasks help users do their jobs. Users are frustrated if they cannot complete a task. Task-oriented topics get the user back "on task." Like a compass on a journey, tasks provide direction.

To make information task oriented, follow these guidelines:

- ❑ **Write for the intended audience.**
- ❑ **Present information from the user's point of view.**
- ❑ **Indicate a practical reason for information.**
- ❑ **Focus on real tasks, not product functions.**
- ❑ **Use headings that reveal the tasks.**
- ❑ **Divide tasks into discrete subtasks.**
- ❑ **Provide clear, step-by-step instructions.**

Write for the intended audience

Before you start writing, be sure that you have a clear understanding of your audience. For example, if you are writing for managers, you might include only high-level tasks, such as evaluating and planning, or a high-level view of other tasks. Similarly, if you are writing for end users, avoid system administrator tasks.

Be sure that the information that you include in your topics is of interest to your audience. For example, your product might have a powerful new help system, but a description of the help system features is of little interest to the person who is installing the product.

The following passage shows a simple task that is explained in detail. However, the audience consists of users who want to use an advanced feature and therefore do not need help performing simple tasks. This information will frustrate all but the most patient advanced user.

Original

To customize your settings:

1. Go to the file tree.
2. Click the **INFODIR** folder.
3. Right-click the SETTINGS.DEF file and select **Edit** from the menu.
4. Change the settings that you want in the file.
5. Click **File** —> **Save** to save the file.
6. Click **File** —> **Close** to close the editor.

Revision

To customize your settings, edit the INFODIR/SETTINGS.DEF file.

In the revision, the task is handled much more simply. The revision quickly provides the users with the information that they need, because the writer understands the skill level of the audience.

For more information about how much detail to provide based on the type of user, see the completeness guideline "Cover each topic in just as much detail as users need" on page 79.

Present information from the user's point of view

Writing from the user's point of view brings the user into the "story," so that users can easily imagine doing the task that you describe. Such writing has these characteristics:

- It is predominantly directed at "you" (second person).
- It uses the active voice, with verbs that denote actions that the user does as opposed to actions that the product does.
- It gives a reason for the actions.

The following passage is written from a remote, impersonal point of view:

Original

> The system should not be shut down during processing. If such a shutdown occurs, the system should be restarted with the START RECOVER command.

Revision

> If you shut down the system during processing, you might lose data. Use the START RECOVER command to restart the system and recover any data from the log.

The original passage is passive and indefinite about who does the action and why. In the revision, the information is presented to make the user an active participant. The phrase "you might lose data" expresses the reason for the action in terms that users can relate to personally—they don't *want* to lose data.

The following passage leaves the users out of the story altogether:

Original

> Subsequent installation of the HIGS feature allows InfoProduct to run unattended.

Revision

> If you want to run InfoProduct unattended, you must install the HIGS feature. You can install the HIGS feature after you install InfoProduct.

Users don't know how the information in the original passage pertains to them. The revision adds the users to the story and explains what they need to *do* to use the feature.

When you provide task help, you have the advantage of knowing where the users are in the product interface and, therefore, what part of the task they need help with. This advantage allows you to write a help topic that's specific to the user's position.

For the following help topic, assume that a user accesses online help from the Primary Key page (which is the third and final page) of the Create Table wizard.

Original

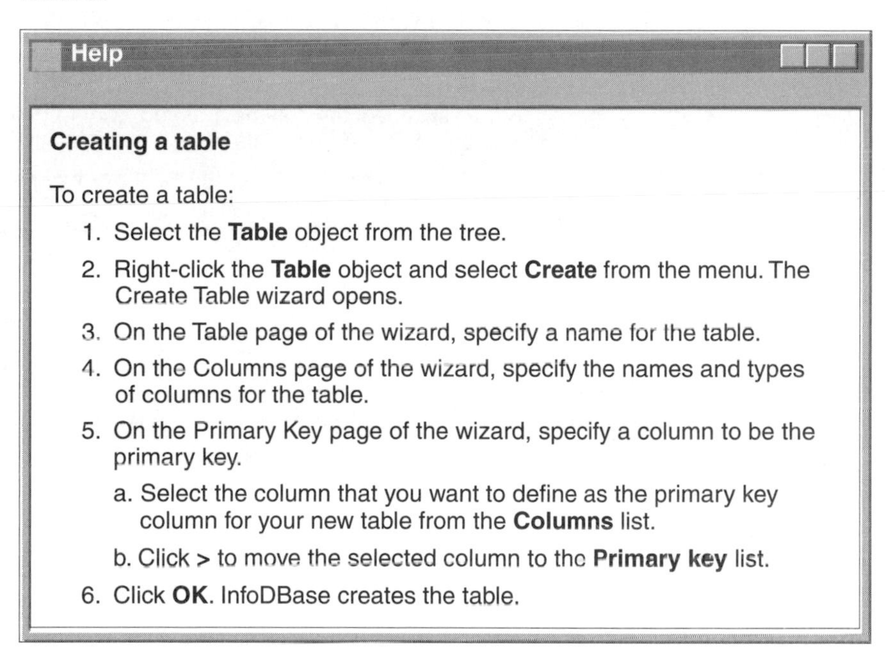

Help

Creating a table

To create a table:

1. Select the **Table** object from the tree.
2. Right-click the **Table** object and select **Create** from the menu. The Create Table wizard opens.
3. On the Table page of the wizard, specify a name for the table.
4. On the Columns page of the wizard, specify the names and types of columns for the table.
5. On the Primary Key page of the wizard, specify a column to be the primary key.
   a. Select the column that you want to define as the primary key column for your new table from the **Columns** list.
   b. Click **>** to move the selected column to the **Primary key** list.
6. Click **OK**. InfoDBase creates the table.

The user has already completed most of the steps of the "Creating a table" task and needs only information about how to specify a primary key, which is a subtask of creating a table. The original help topic wastes the user's time because it provides steps that the user has already completed.

Revision

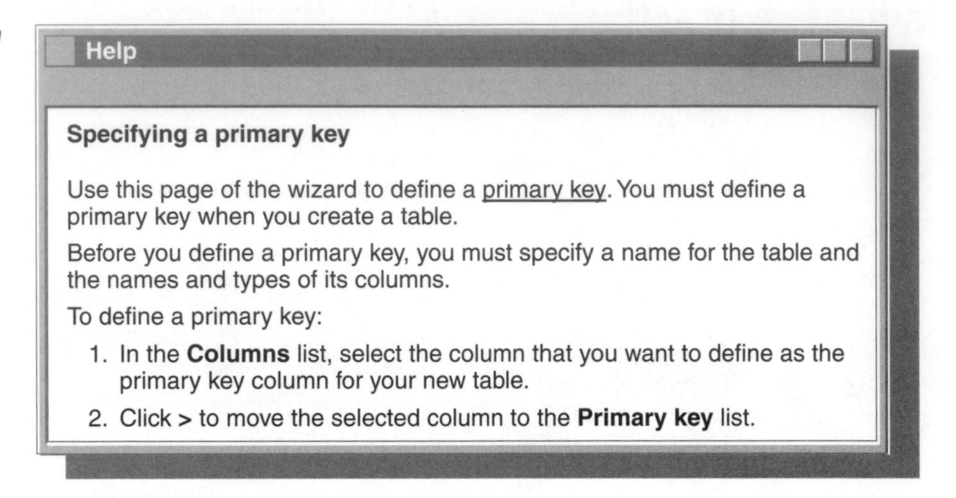

The revised help topic presents the task from the user's probable position. The revised topic also provides a link to a definition of a primary key for users who need to understand that concept before proceeding with the task.

When you understand the user's point of view, you can write task-oriented information that helps users do their tasks.

Indicate a practical reason for information

Giving users the information that they need is only part of task-oriented writing. Users need a practical reason for the information. They need to understand *why* you are giving it to them—how it is relevant to their task. The goal that the information serves must be apparent. A task is not just an activity, but an activity that is directed to a particular end.

To ensure that the information that you provide is relevant to the task, follow these guidelines:

- ❑ Relate details to a task where appropriate.
- ❑ Provide only a necessary amount of conceptual information in task topics.

Relate details to a task where appropriate

In a task topic, users should never wonder, "But why are you telling me this?" For example, to state that the records in a file have a certain size might leave users wondering, "So what?" However, if you tell them that they must build a library to hold the file and that the record size affects the way that they do this, they can understand why you are telling them about record size.

In a task topic, facts can puzzle users if you don't indicate what significance the facts have, as shown in the following passage:

Original

> If the NORES option is used, the routines are link-edited as part of the load module. If the RES option is used, the routines are loaded separately.

Revision

> Use the NORES option when you have sufficient space for routines to be link-edited as part of your load module. Use the RES option to save space by loading the routines only when you need them.

The original passage explains what the options do, but it does not relate that information to the user's task of deciding which option to use. In the revision, the facts are restated so that the user understands when to use which option.

At first glance, the following sentence appears to be only descriptive and to have no practical application:

Original

> The BW_Message mapping table in the Data_LM directory can contain warning messages that are issued by InfoProduct when you create a request.

Revision

> After you create a request, check the BW_Message mapping table to see if InfoProduct issued any warning messages. The BW_Message mapping table is in the Data_LM directory.

Users might glance at the original sentence and move on because it doesn't seem relevant. The revised sentence relates the information to the task of creating a request, so that users understand why the information is significant.

Provide only a necessary amount of conceptual information in task topics

Task topics should focus on providing only the information that supports the task. Task topics can contain:

- Rationale for doing the task
- Requirements for doing the task (such as prerequisite tasks or software)
- Steps of the task
- Tips and examples to support the steps
- Information about follow-on (postrequisite) tasks

Do not mix significant amounts of conceptual information with task information. Users want to learn to do things without needing to learn all the concepts that are associated with a product. In task topics, provide steps as early as possible, and explain minor concepts briefly, if at all.

Separate the explanations of major concepts from the task topic by creating a topic for each major concept. Provide links to those concept topics from your task topic. That way, users who already understand the concepts can do the task, and users who need to learn about the concepts can go to the concept topics.

This guideline deals with overcompleteness and is similar to the completeness guideline "Include only necessary information" on page 84.

The following topic explains the concepts of a project and a suite before presenting the task:

Original

Creating test cases

Projects are collections of files that are related to a test case. When you create a test case, you must also create a project. You can, however, create a project before you create a test case. Each project can contain multiple test case files but only one main test case. You create projects in suites. Each suite can contain multiple projects. If you create one suite for each function that you test, the projects in a suite can contain the test cases that you create to test the function.

To create a suite:

1. Right-click a suite object and select **New** from the menu.
2. Specify details in the New Suite window.

To create a project:

1. Right-click a suite object and select **New Project** from the menu.
2. Specify details in the New Project window.

To create a test case:

1. Right-click a project and select **New Test Case** from the menu.
2. Specify details in the New Test Case window.
3. Click **Record** and work with the product that you want to test.
4. Click **Save** after you finish recording your actions.

The original topic starts with conceptual information and then tells the user how to create the needed elements. Users need to read through a big paragraph of information before they can start the task.

Revision

Creating test cases

You can create one or more test cases for each function that you want to test.

Before you begin:

1. Create a suite for the function that you are testing.
2. Create a project within the suite to hold the test cases and files.

To create a test case:

1. Right-click a project and select **New Test Case** from the menu.
2. Specify details in the New Test Case window.
3. Click **Record** and work with the product that you want to test.
4. Click **Save** after you finish recording your actions.

Related topics
Creating suites
Creating projects
Suites
Projects
Test cases

The revised topic does not explain all of the relationships of the elements. These elements are explained in detail in concept topics. Links are provided from the bottom of the task topic to these concept topics. In addition, the revised topic uses a prerequisite section that is labeled "Before you begin" to show that two of the elements must be created before you begin this task.

This separation of task topics and concept topics allows the writer to share the same concept topics across multiple tasks. For example, the task topic for creating a project can also provide a link to the "Projects" concept topic.

The revised task does not provide the substeps of the prerequisite tasks. Those steps are provided in separate tasks that are linked to from this task. Each of the three tasks is handled in a separate topic as explained in the guideline "Divide tasks into discrete subtasks" on page 33 of this chapter.

Focus on real tasks, not product functions

A *real task* is a task that users want to perform, regardless of whether they are using your product to do it. Tasks that are imposed by the product are *artificial tasks*. It is all too easy in technical writing to lose sight of the real tasks and get caught up in the tasks that are dictated by the product. When you live and breathe a product for months, you might forget that the user's tasks and the product's tasks are not necessarily the same.

Examples of real and artificial tasks are:

- Users want to edit a table, but the writer introduces this task as "using the table editor" instead of "editing a table."
- Users want to count the records in a file, but the writer introduces the task as "using the CNTREC utility" instead of "counting records with the CNTREC utility."

Sometimes the design of a product forces you to write about artificial tasks. For example, users want to create a graphic, but they must first set up a container to hold sets of graphics. In such cases, you can focus on the task instead of the design by writing about "organizing your graphics with containers" rather than "creating containers."

In the following lead-in to the steps, the writer makes the common mistake of describing the task in product-specific terms:

Original

To use the InfoInstaller utility:

1. Open the InfoInstaller window by typing `infoinst` at the command line.
2. Complete the InfoInstaller window by specifying the installation parameters.
3. Click **OK**. InfoInstaller installs InfoProduct.

The original introduction assumes that users understand the task in terms of the tool that they need to use to do the task. Although some users might know what InfoInstaller is, all users know what installation is.

Revision

To install InfoProduct:

1. Type `infoinst` at the command line. The InfoInstaller window opens.
2. Specify the installation parameters in the window.
3. Click **OK**. InfoProduct is installed.

The revised introduction and steps separate the task from the tool so that users can relate to the real task of installing the product instead of to the tool that they use to do so.

The following introduction shows a topic that explains how to use the product rather than how to perform real tasks:

Original

> This topic explains how to use the following menu choices under **File**:
>
> **Open** Opens an existing file.
> **New** Creates a file.
> **Save as** Saves to a new file with a different name.

The original text assumes that the user is examining the interface and wondering what each menu item does. This type of information is appropriate for *contextual help* (help that is relevant to where a user is in a product, such as help for the selected control), but not for a task topic. Users want information about how to do real tasks, not a list of the buttons and fields in the product interface.

Revision

> This topic explains how to work with a document. You can do the following tasks:
>
> ❑ Create a document
> ❑ Open an existing document
> ❑ Rename a document

The revision shows information that is presented in terms of real tasks.

By presenting the information from the perspective of tasks that users recognize and want to perform, you make the information more relevant to users.

The following topic presents the task in terms of the window that is used to do the task:

Original

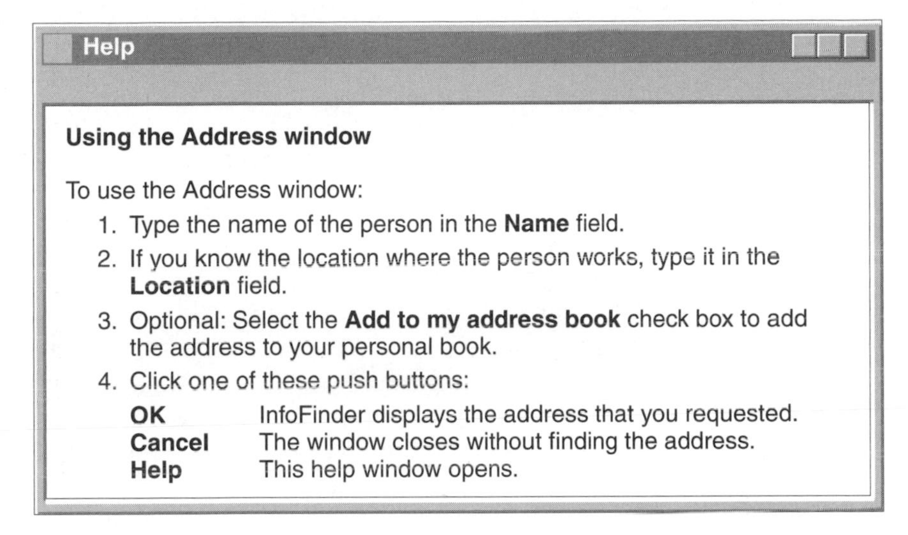

The original help topic tells the user how to use the window; it explains how to complete the fields in the window, but it never explains what real task the user is doing. The original help topic also contains extraneous information about the window that isn't pertinent to the real task.

Revision

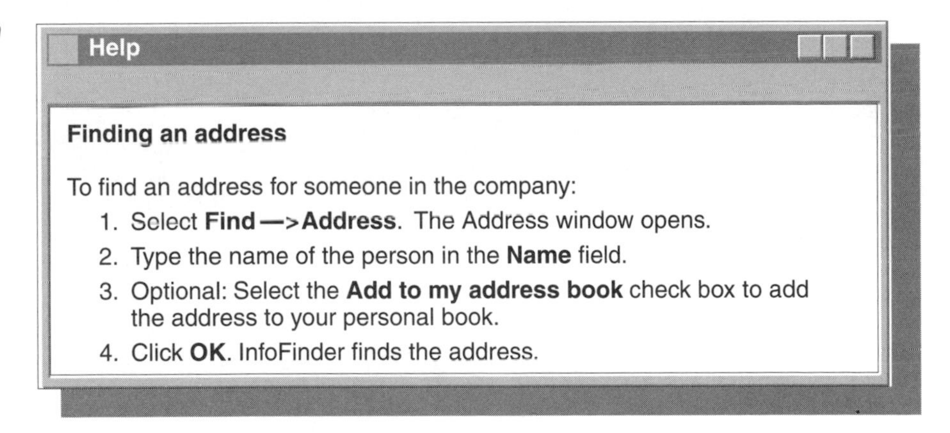

The revised help topic presents the information in terms of the task that the user wants to do and the steps that are involved in the task.

Keep task steps focused on the real task. Avoid littering steps with feature "advertisements" unless the features are especially helpful to the user as part of the task. Apply the completeness guideline "Cover each topic in just as much detail as users need" on page 79.

The following help topic contains unnecessary information about features that is mixed in with the task:

Original

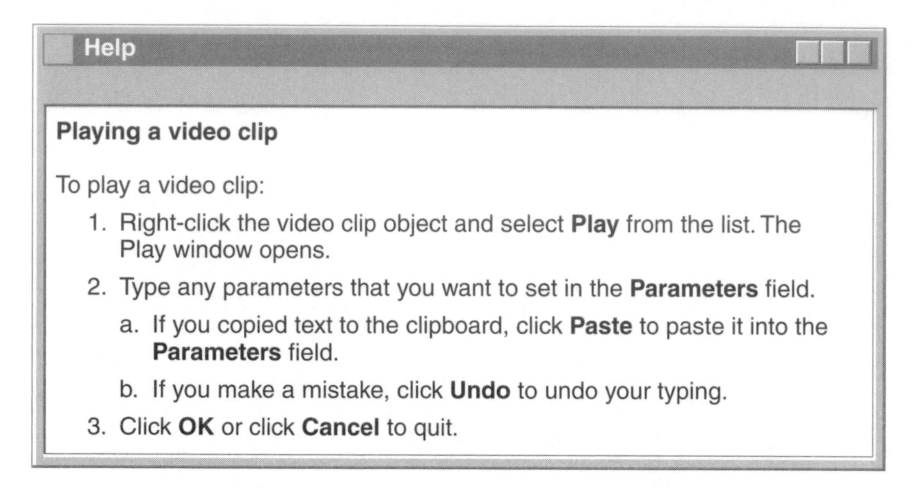

Revision

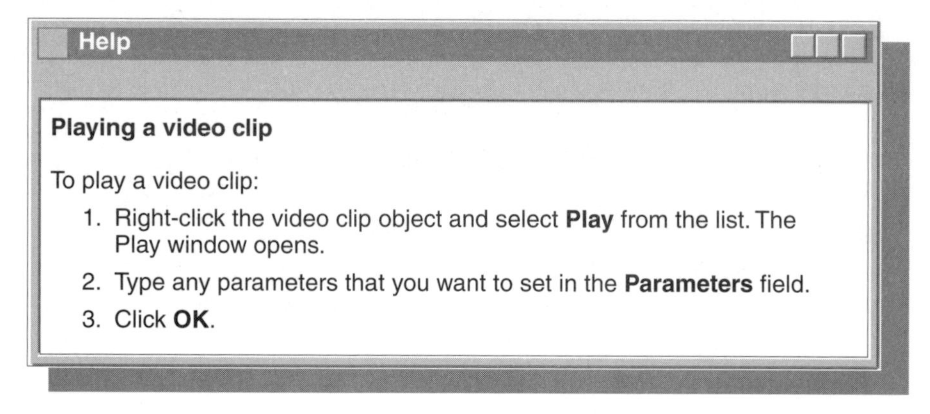

Steps 2a and 2b in the original help topic are *not* part of the task. Clicking **Cancel** is also not part of the task. These are only features that users can use. If you document every feature like this, your tasks will become unwieldy and lose their focus. The revision shows the real steps of the task.

Always focus on real tasks, not artificial tasks, when you write task topics, headings, and steps.

Use headings that reveal the tasks

The first place to tell users why they are being given information is in the heading. Users should get an accurate idea about the content of the topic from the heading. Each heading should reveal that the topic contains information about a task, and what that task is.

A static heading like "Authorizations" might be fine for a concept or reference topic, but for a task topic, the heading should reveal the task that users are being told how to do. A more helpful heading might be "Authorizing users" or "Setting authorizations for a user ID."

Be careful not to mislead users by using *pseudo-task* headings. Pseudo-task headings start with vague verbs, such as "understanding" and "learning" (and sometimes "using"). Pseudo-task headings mislead users in the following ways:

- Pseudo-task headings make a topic appear to be task-oriented when it is actually full of conceptual or reference information. For example, a topic called "Understanding the file system" probably does not contain steps for understanding the file system, but instead contains reference information about the file system. A more appropriate heading might simply be "The file system."
- Pseudo-task headings substitute artificial (product-imposed) task headings in place of real task headings. For example, a heading like "Using the SpellMaster tool" is hiding the real task, and should probably be called "Checking the spelling in a document" instead. The guideline "Focus on real tasks, not product functions" on page 27 of this chapter provides more discussion and examples of real and artificial tasks.

Headings are elements that help users find information. Good headings produce a good table of contents, from which users can predict what they will find in each topic: task, conceptual, or reference information. Therefore, this guideline is related to retrievability as well as task orientation.

The following headings are misleading and unhelpful:

Original

Register usage
Administering authorization
The Dial-up function
Session initialization
Using the Define Font Window
Understanding hardware requirements

Some of the original headings hide the fact that the topic contains task information; others define the tasks in terms of the product; and the last heading misrepresents a reference topic as a task.

Revision

```
Linking with registers
Authorizing access to data
Dialing up the computer
Initializing a session
Defining a font
Hardware requirements
```

The revised headings clearly reveal the type of information that each topic contains; they use verbs that express user actions, and they indicate the goal of the action.

The revised headings use gerunds for task-oriented headings. Gerunds are not the only alternative for task-oriented headings. Your style guideline might call for headings like "How to define a font" or "Steps for defining a font" instead.

The following excerpt from a list of topics doesn't give the user any idea of what task the information supports:

Original

```
Working with InfoDataMagic
  Using collectors
    Starting collectors
    Configuring collectors
    Stopping collectors
  Using repositories
```

Revision

```
Managing copied data
  Collecting data
    Starting a data collection
    Configuring collection properties
    Stopping a data collection
  Storing data
```

In the original list of topics, the headings could apply to a variety of tasks. Users who are not familiar with the terminology for the product have no idea what tasks the topics explain and could easily miss these sections in a list of topics. The revised list of topics shows task-oriented headings that should make sense to all users.

Divide tasks into discrete subtasks

After you identify your main tasks, you need to divide them into discrete subtasks so that you can provide usable step-level information. If you tried to provide step-level information for the main task of writing a news story, you would have thousands of steps. Instead, you divide the task of writing a news story into its main subtasks, for example, researching the material, drafting an outline, writing a first draft, and revising the draft. Repeat the process of dividing tasks until you have groups of tasks for which you can provide step-level information.

Keep in mind that the number of steps in each task and subtask should be nine or fewer, as discussed in the clarity guideline "Keep elements short" on page 124. If you find that a task takes more than nine steps to do, determine whether the task can be divided again or whether some steps can be nested under others. For more information about nesting steps, see the guideline "Group steps for usability" on page 38 of this chapter.

Before you write task information, you need to know how the task is related to other tasks. You need to understand the task sequences, levels of tasks, and interdependencies. Is the task really a discrete task? Does it have subordinate tasks? Is it a subtask of a larger task? Is it a subtask of more than one task?

Although this guideline applies to task orientation, it does overlap with the following organization guidelines: "Organize information into discrete topics by type" on page 218 and "Organize tasks by order of use" on page 222.

The following task overview is organized according to the order that the user should do the subtasks. You can easily replace this overview with a high-level task that lists appropriate subtasks.

Original

To migrate to version 8, you need to read these topics:

- ❑ Hardware requirements
- ❑ Software requirements
- ❑ Applying the latest updates
- ❑ Stopping InfoProduct processes
- ❑ Backing up your infoproduct.inf file
- ❑ Running the migration utility for Windows
- ❑ Running the migration utility for UNIX
- ❑ Verifying migration
- ❑ Setting up a new profile

Revision

To migrate to version 8:

1. Ensure that you have the correct hardware and software:
   - ❑ Hardware requirements
   - ❑ Software requirements
2. Apply the latest updates.
3. Stop all InfoProduct processes.
4. Back up your infoproduct.inf file.
5. Run the migration utility.
   - ❑ On Windows
   - ❑ On UNIX
6. Verify the migration.
7. Set up a new profile.

After you finish migrating InfoProduct, you can start the processes again.

The revision shows the relationship of the subtasks to each other and to the higher-level task. The revision also mentions what to do after completing the task.

The following list of topics shows a hierarchy of tasks in which the task of verifying the installation is combined with the task of configuring parameters:

Original

Installing InfoProduct

- Checking software
  - ❑ Linux
  - ❑ Windows
- Starting installation
  - ❑ Linux
  - ❑ Windows
- Configuring parameters and verifying installation
  - ❑ Linux
  - ❑ Windows

In the original list, the task of verifying installation is repeated for both Linux and Windows operating systems. The task of verifying installation is not part of the task of configuring parameters. One user might verify the installation while another user configures the parameters. The two tasks should not be combined.

Revision

Installing InfoProduct
- Checking software
  - Linux
  - Windows
- Starting installation
  - Linux
  - Windows
- Configuring parameters
  - Linux
  - Windows
- Verifying installation

The revision shows verifying installation as the last, discrete task at the same level as the other main subtasks of installing the product. The revision allows for a logical separation of different tasks.

Consider the case in which a user might need to configure additional parameters at some time other than during installation. Seeing the steps for verifying installation with the steps for configuring parameters does not make sense to that user. In addition, if the steps for verifying installation are the same on Linux as those on Windows, you can write the information once instead of repeating those steps in the topic for Linux and the topic for Windows.

Be sure to separate your tasks into discrete subtasks such that each subtask contains only one distinct task, yet contains sufficient content to stand alone.

Provide clear, step-by-step instructions

Steps make up most tasks. Occasionally a task has only one step and can be described in a paragraph, but most tasks are performed as a series of ordered steps. Task orientation extends to the level of the lowest step. Any step that is not clearly written or not ordered correctly can cause your users to make mistakes and be unable to do a task.

When you write task information, you are usually in a position to notice usability problems in time to suggest product improvements. For example, you might identify cumbersome steps that can be streamlined or avoided, or steps where users are repeating actions that they've already performed (such as typing a long serial number in more than one window). If you have difficulty documenting a task, consider whether there might be a problem with the way that the product works. Keep the needs and interests of your users in mind as you write; there is no such thing as a product that is too usable. In fact, the more usable a product is, the less users must rely on low-level step information.

Take the time to organize your steps from your users' perspective. Know the answers to these questions: How do tasks relate to each other? When are steps subtasks? What constitutes a step? Where does a step start and end? Are some steps subordinate to others? Which steps are optional? Which steps are conditional?

The following guidelines can help you provide clear step-by-step instructions:

- ❑ Make each step a clear action for users to take.
- ❑ Group steps for usability.
- ❑ Clearly identify optional steps.
- ❑ Identify criteria at the beginning of conditional steps.

Make each step a clear action for users to take

Each step should correspond to an action (high level or low level) that the user performs. Tasks do not discuss how the user and product interact; tasks list only the actions that the users do to complete their task.

A step isn't complete unless it has an action for the user to do. One of the following steps has no user action:

Original

3. Click **OK**.
4. The installation begins.
5. After installation completes, restart your system.

Revision

3. Click **OK**. The installation begins.
4. After installation completes, restart your system.

In the original set of steps, step 4 describes a product action, not a user action. In the revision, step 4 is combined with step 3 because it is the result of step 3, not a separate step.

To ensure that each step gives clear direction to users, include an imperative verb (a verb that instructs the user to take an action) in the first sentence of every step. When you make style decisions for your product, you might pick a specific way to phrase step-level information. For example, you might choose to "place" your user before stating the user action, as in "In the first column of the table, type the date." Alternatively, you might choose to put the user action before the placement, as in "Type the date in the first column of the table." Both approaches include the imperative verb in the first sentence of the step.

Some decisions are trickier than others. Consider the following set of steps:

Original

3. Click **OK**.
4. The InfoUpdater should stop.
   - If it doesn't stop, repeat steps 2 and 3.
   - If it does stop, go to step 5.
5. Run the InfoVerify tool to check for viruses.

Revision

3. Click **OK**. InfoUpdater should stop.
4. If the InfoUpdater is not stopped, repeat steps 2 and 3.
5. Run the InfoVerify tool to check for viruses.

Step 4 of the original set of steps breaks the rule of including an imperative verb in the first sentence of every step. Step 4 is not a step. In the revision, step 4 includes an imperative verb and is easier for users to follow.

Group steps for usability

Group steps to help users relate to the task. If you instruct users to do one action or click after another, the task can become mind numbing for the user. If you can group minor steps together into a larger step, users can think of the steps in relation to the goal of completing the task.

For example, instead of interpreting the steps as "First I click here, then I fill out that field, then I click over there, then I choose a button here," users might be able to think of the steps in terms of "First I set my preferences, then I specify the server information." In this way, you not only help users relate to the steps that they are doing, you also streamline the steps and make each step easier for users to find.

The following steps are in the correct order, but they don't correspond to the way that the user thinks about the task:

Original

To add a setting to your profile:

1. Select the profile object that you want and right-click.
2. Select **Properties** from the menu.
3. In the Properties window, find the name and path of the profile file.
4. Close the Properties window.
5. Open your profile file in a text editor.
6. Add the setting to your profile file in the settings section.
7. Save the profile file.
8. Run the profile command with the -file *YourProfileName* option.

The original set of steps gives each step the same weight. It treats trivial steps, such as closing a window, the same way that it treats more significant steps, such as running the command. The original set of steps ignores the relationship of each step to its surrounding steps.

Revision

To add a setting to your profile:

1. Determine the name of the profile file that you want to add the setting to:
   a. Right-click the profile object that you want and select **Properties** from the menu.
   b. In the Properties window, find the name and path of the profile file.
2. Update the profile file with the new setting:
   a. Open your profile file in a text editor.
   b. Add the setting to your profile file in the settings section.
   c. Save the profile file.
3. Run the profile command with the -file *YourProfileName* option.

The revised set of steps shows the relationships of some of the steps to each other and shows how they make up the two higher-level steps: finding the name of the file and updating the file. The revised set of steps also downplays some of the trivial steps by merging them or omitting them. Where the original set of steps shows a linear progression of one action or click after another, the revised set of steps shows what each step accomplishes toward completing the whole task. Also, by combining the steps into higher-level steps, the revision minimizes the number of steps.

Unordered lists, or bulleted lists, provide another way to subordinate information or actions in steps. Be sure to use unordered lists only for tasks that are not sequential. You could use an unordered list to show, for example, more than one way to do a step.

The following set of steps uses unordered lists to try to get both a new user and a returning user through the first two windows needed for a task:

Original

To purchase tickets online:

1. Go to www.e-infoticket.com.
2. Select one of the choices.
   - If you are a new user, click **Register**.
   - If you registered before, click **I Have an Account**.
3. Complete the fields on the form.
   - If you are a new user, the form requires a credit card number.
   - If you are a returning user, specify your password.
4. Click **Select Tickets to Purchase**.
5. Choose the tickets that you want.
6. Click **Submit** and wait a few seconds for a confirmation message.
   - If a message tells you that the tickets are purchased, write down or print the confirmation number. You have finished.
   - If no message appears, repeat steps 5 and 6.

In the original set of steps, both the new user and existing user need to read both step 2 and step 3 to figure out what to do. The original steps are focused on the product, not the flow of the task.

Revision

To purchase tickets online:

1. Go to www.e-infoticket.com.
2. If you are a new user, set up an account:
   a. Click **Register**.
   b. Specify your credit card number and a name and password for your account.
   c. Click **Submit**. When the account is set up, a message will tell you to proceed.
3. Click **I Have an Account**.
4. Specify your name and password.
5. Click **Select Tickets to Purchase**.
6. Choose the tickets that you want.
7. Click **Submit** and wait a few seconds for a confirmation message.
   - If a message tells you that the tickets are purchased, write down or print the confirmation number. You have finished.
   - If no message appears, repeat steps 6 and 7.

In the revision, step 2 is used to prepare the new user for the remaining steps. Thus both types of users can follow one clear set of steps.

When you use sublists to group steps, be sure to use the right type of sublist for the situation. The following example shows a step divided into substeps:

Original

3. Copy the contents of the file system from the source disk to the target disk. The steps that you use depend on the location of the target disk:

   a. If the target disk is on the same computer as the source disk, run the copyfilesystem command.

   b. If the target disk will replace the source disk, follow these steps:

      1. Copy the contents of the source disk to tape.
      2. Replace the source disk with the target disk.
      3. Configure the target disk.
      4. Copy the tape contents to the target disk.

In the original step, ordered substeps are used to show two choices that are mutually exclusive. Because the choices are not meant to be performed in order, the substeps are misleading.

Revision

3. Copy the contents of the file system from the source disk to the target disk. The steps that you use depend on the location of the target disk:

   - If the target disk is on the same computer as the source disk, run the copyfilesystem command.
   - If the target disk will replace the source disk, follow these steps:

      a. Copy the contents of the source disk to tape.
      b. Replace the source disk with the target disk.
      c. Configure the target disk.
      d. Copy the tape contents to the target disk.

In the revision, the substeps are replaced with bulleted options. Users follow the instructions in one bullet or the other.

Clearly identify optional steps

Optional steps are steps that a user can skip and still complete the task successfully. As mentioned in the guideline "Focus on real tasks, not product functions" on page 27 of this chapter, try to keep your tasks free of feature clutter by eliminating steps that are superfluous to the task. However, in cases where optional steps support the task, include them in the task, but identify them as optional. For example: "2. Optional: Define a profile for your startup parameters."

The following list of steps identifies step 1 as optional, but not step 2:

Original

1. Optional: Click **Timeout Settings** to specify timing settings for the profile. The Timeout Settings window opens.
2. In the Timeout Settings window, specify the number of seconds before the system is to restart.

The original set of steps is confusing. Users who choose to skip step 1 are unable to do step 2. Because step 2 can be done only if the user follows step 1, step 2 must also be optional.

First revision

1. Optional: Click **Timeout Settings** to specify timing settings for the profile. The Timeout Settings window opens.
2. Optional: In the Timeout Settings window, specify the number of seconds before the system is to restart.

The first revision shows both steps as optional. However, the first revision is still not logical because users who choose not to perform step 1 cannot perform step 2. So step 2 is not optional by itself.

Second revision

1. Optional: Specify timing settings for the profile:

   a. Click **Timeout Settings** to open the Timeout Settings window.
   b. Specify the number of seconds before the system is to restart.

The second revision shows step 1 as an optional step that consists of two substeps. If users choose to follow step 1, they must perform both steps 1a and 1b, which is the only combination that makes sense.

Take care to clearly identify optional steps. Use "Optional" for optional steps. However, do not follow the word "Optional" with the phrase "If you want to" or "You can" because these phrases are redundant with the word "Optional."

Identify criteria at the beginning of conditional steps

Conditional steps are those that users follow only if certain criteria apply. Conditional steps generally begin with the word "If," as in, "If you run test cases in batch mode, complete the fields on the Batch page." Users who meet the criteria for the step must follow the step. Always start conditional steps with the condition. That way, users who do not meet the criteria can skip the step after reading the condition.

Although the following steps are conditional, users might find themselves halfway through the steps before they realize that they don't need to do them.

Original

1. Register your computer as a client if you are not yet registered on the LAN.
2. In the **Number** field, specify your 12-digit serial number if your software is not yet registered.
3. Run the InfoExec program to reconfigure your settings. (InfoExtended only)

Because users rarely read ahead when following steps, the original steps might cause some users to take links that do not apply to them, start typing serial numbers, or try to run programs that they don't need.

Revision

1. If you are not yet registered on the LAN, register your computer as a client.
2. If your software is not yet registered, in the **Number** field, specify your 12-digit serial number.
3. InfoExtended only: Run the InfoExec program to reconfigure your settings.

In the revised steps, the condition for each step is stated before the action. Users who are registered on the LAN can skip step 1, users who registered their software can skip step 2, and users who are not using InfoExtended can skip step 3.

The following step is introduced in a potentially confusing way:

Original

3. To specify the date parameter, click **New**.

Users might read the original step as optional, required, or conditional. The revisions show more specific phrasing for all three situations.

Revision: optional

3. Optional: Click **New** to specify the date parameter.

Revision: required

3. Click **New** to specify the date parameter.

Revision: conditional

3. If your date parameter is not defined, click **New** to specify it.

Take care to clearly identify conditional steps. Use "If" and state conditions early for steps that do not apply in all situations.

In sum

Task-oriented information focuses on the user's tasks and is presented from the user's perspective. Divide task information into tasks and subtasks. Provide steps that are clear, imperative, and grouped for usability. Do not clutter task topics with conceptual information. Get your users "on task" as quickly as possible.

This chapter provides guidelines to help you ensure that your topics are task oriented. Refer to the examples in the chapter for practical applications of these guidelines.

When you review technical information for task orientation, you can use the checklist on page 46 in two ways:

- ❑ As a reminder of what to look for, to ensure a thorough review
- ❑ As an evaluation tool, to determine the quality of the information

Based on the number and severity of items that you find, decide how the information rates on each guideline for this quality characteristic. You can then add your findings to "Quality checklist" on page 387, which covers all the quality characteristics.

Although the guidelines are intended to cover all areas for this quality characteristic, you might find additional items to add to the list for a guideline.

Guidelines for task orientation	Items to look for	Quality rating
Write for the intended audience.	• Audience is clearly defined. • Tasks are appropriate for the intended audience.	1 2 3 4 5
Present information from the user's point of view.	• Information is directed at the user. • Information denotes what the user does, as opposed to what the product does. • Contextual help reflects the user's position.	1 2 3 4 5
Indicate a practical reason for information.	• Details relate to tasks. • Conceptual information supports the task. • Only a necessary amount of conceptual information appears before a task.	1 2 3 4 5
Focus on real tasks, not product functions.	• Information focuses on real tasks, not artificial tasks. • Focus is on tasks, not features and interface elements.	1 2 3 4 5
Use headings that reveal the tasks.	• Headings reveal tasks. • Headings do not contain pseudo tasks.	1 2 3 4 5
Divide tasks into discrete subtasks	• Relationship of tasks to subtasks is clear. • Subtasks are discrete.	1 2 3 4 5
Provide clear, step-by-step instructions	• The first sentence of each step includes an imperative verb. • Steps are weighted appropriately. • Substeps and unordered lists are used appropriately. • Optional steps are clearly marked. • Conditional steps begin with the condition.	1 2 3 4 5

Note: The scale for the quality rating goes from very satisfied (1) to very dissatisfied (5).

Chapter 3

Accuracy

Be sure you are right, then go ahead. —David Crockett

As a writer of technical information, you have an important responsibility: to provide information that is accurate and free of errors. Users depend on you for that accuracy; they stake their time and money on it. Achieving accuracy in technical information is to a writer as finding the bull's-eye of a target is to an archer; it is an all-important goal, albeit a goal that isn't always easy to achieve.

For technical information to be accurate, every piece of information must be accurate, including conceptual information, factual statements, procedures, graphical elements, and other details in the writing.

Inaccuracies might or might not be obvious. Those that aren't obvious are often more significant, because the user might take action based on inaccurate information.

For example, imagine someone leaving a message on your answering machine with directions to "turn right at the dead end." Before you get in your car, you know that this is incorrect because you can't turn right at a dead end. Therefore you take steps to resolve the inaccuracy, like looking at a map or asking for clarification.

If the person's message says "turn right at Front Street and go straight until you cross the railroad tracks," you could drive for quite some time before realizing that no railroad tracks are in this direction, and that you were actually supposed to turn left instead of right. You'll probably be more upset by the second set of directions than the first.

In the article "Docs in the Real World," Carolyn Snyder and Jared Spool recognize the vital importance of information accuracy when they coin the term "trust-breakers." These are errors in information that cause users to avoid the information because they don't trust it. Snyder and Spool make several observations about the effects of inaccuracies on users:

- Even small inaccuracies damage the confidence of some users, especially those users who don't know how to work around the problem. When they discover an error, they tend to give up on the information and call the help desk.
- The information often takes the blame for inaccuracies in the underlying product or in related products. Users don't care whose fault an error is; they are simply less willing to use the information in the future.
- After experiencing major problems in information, users are less likely to give the information another try, even in future releases.

Regardless of how much time users take to detect an inaccuracy, their confidence in the information is eroded; they might even choose not to use the information or the product at all.

To make information accurate, follow these guidelines:

- **Write information only when you understand it, and then verify it.**
- **Keep up with technical changes.**
- **Maintain consistency of all information about a subject.**
- **Use tools that automate checking for accuracy.**
- **Check the accuracy of references to related information.**

Write information only when you understand it, and then verify it

By the time you publish information, you must understand the subject well enough that you can tell users what they need to know. Sometimes you can understand the subject quickly and be able to write about it accurately and clearly without much difficulty. At other times, the subject is new and not well documented.

Perhaps a software product is not available for you to try, or technical experts are unable to explain the subject to you at the beginning of the writing cycle. When you can't get first-hand experience with a product or can't get help from technical experts, you must start with what you do understand to write your first draft.

Consider including questions to the experts in your early drafts to solicit the explanations that you need to fine-tune the information for users. Technical experts are sometimes better at reviewing something that is written and identifying the inaccuracies than they are at documenting a technical subject from scratch. Write the information and have it reviewed as many times as necessary to ensure that it is accurate by the time when you publish the information.

Although you do need to understand the technical subject that you write about, you don't need to understand it as well as the developers. In fact, if you understand the internal workings of the product too well, you run the risk of including more information than users need. (Chapter 4, "Completeness," on page 75 provides guidelines for knowing how much information is needed.)

If you understand what you're writing about, you'll be better able to explain concepts and define terms, as in the following explanation of a wide area network:

Original

> A wide area network (WAN) is a network that encompasses a wide area.

Revision

> A wide area network (WAN) is a computer network that encompasses a large geographical area. A WAN usually includes two or more local-area networks (LANs).

In the original passage, the writer does not yet understand the concept of a WAN, and the resulting definition is too simplistic to be helpful to most users. Perhaps the original passage is a first attempt at writing the definition. After seeking a technical review, the writer is able to improve the accuracy of the definition as shown in the revised passage.

When you write procedures, you need to understand how users are likely to use the procedures to perform relevant tasks. If you don't understand how users perform the task, the task orientation of the information will suffer, and you might also end up with an inaccurate procedure. For example, you might obtain a general outline of the procedure from a developer who is not thinking about how users will use the procedure.

The following topic shows a procedure on the Web that does not satisfy all users:

Original

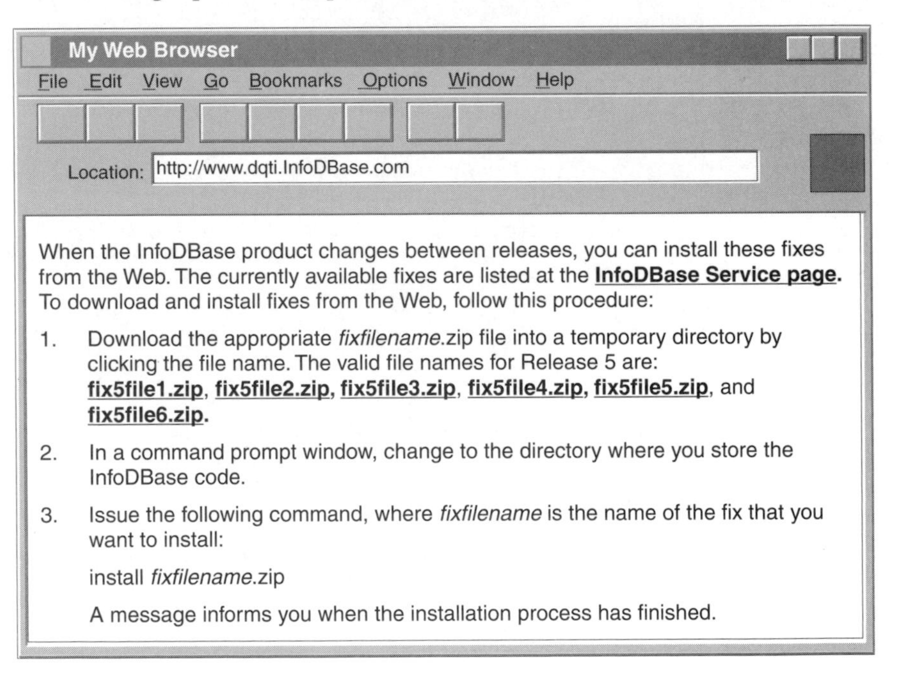

When the InfoDBase product changes between releases, you can install these fixes from the Web. The currently available fixes are listed at the **InfoDBase Service page.** To download and install fixes from the Web, follow this procedure:

1. Download the appropriate *fixfilename*.zip file into a temporary directory by clicking the file name. The valid file names for Release 5 are: **fix5file1.zip**, **fix5file2.zip**, **fix5file3.zip**, **fix5file4.zip**, **fix5file5.zip**, and **fix5file6.zip**.
2. In a command prompt window, change to the directory where you store the InfoDBase code.
3. Issue the following command, where *fixfilename* is the name of the fix that you want to install:

   install *fixfilename*.zip

   A message informs you when the installation process has finished.

The original procedure has several problems:

- The first paragraph states that the fixes are available on the Service Web page but does not instruct the user to follow the link. Many users focus on the actual procedure and might never follow the link to the Service Web page.

- Step 1 specifies the names of the fix files. The writer, perhaps with good intention, lets users double-click the file names directly, even though this list is for only one release. Users of other releases cannot accomplish the task of downloading fixes by using this procedure.
- The download list in step 1 is likely to become out of date if the writer doesn't update the procedure each time that a new fix becomes available.

Revision

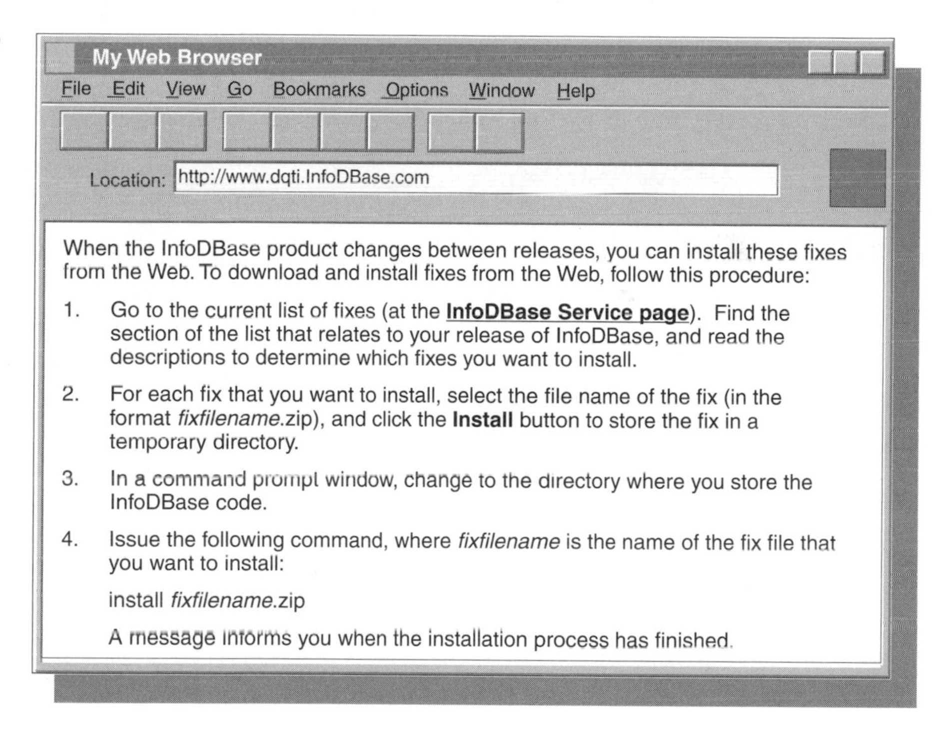

The revised procedure accurately takes into account how users will install fixes. Step 1 instructs the user to go to the Service Web page to identify the appropriate fixes and to download and install them. The revised procedure does not mention specific file names and is therefore less likely to become outdated.

The best way to understand a technical subject that you are writing about is to use the product as a user would. The more that you, as a writer, can use a product, tool, or interface, the more responsible you can be for the accuracy of the information. For example, when you write a procedure for creating a database view that is based on two tables, you can use the interface to decide what steps to include. Then, after you finish writing the procedure, you can test it; if you successfully create the view, you know that the procedure is accurate.

When time and circumstances allow, try to ensure the accuracy of your information in one or more of the following ways:

- Schedule an informal usability walkthrough in which you and other writers use the information to perform some or all of the important user tasks.
- Observe a formal usability walkthrough, in which actual users do certain tasks with the information and associated product, tool, or interface.
- Visit one or more of the users of your information to observe them as they do their day-to-day tasks.

These kinds of activities increase your knowledge of how your users do their jobs so that you can develop accurate information that makes their jobs easier.

Even writers who use the product, tool, or interface extensively and participate in walkthrough activities must also seek feedback from others:

- Technical reviewers review and validate the accuracy of technical information.
- The writing team reviews and validates the accuracy of nontechnical information such as boilerplate text.

You can solicit this feedback informally, while you write small parts of your information, as well as formally, when you hold reviews or inspections.

Before writing technical information, make sure that you understand the subject well enough so that you can explain to users what they need to know. Whenever possible, use the product, tool, or interface as you are developing the information, and always validate the accuracy before delivering the information to users.

Keep up with technical changes

Technical information that is not current is inaccurate. Information can become outdated from one release to the next or within a development cycle.

Each time that you publish a new edition of your information, ensure that trademarks, product names and release levels, and boilerplate information are current. Try to avoid mentioning specific release levels of products unless you have a valid technical reason to do so, as in the following cases:

- Marketing information that highlights the features of a new release or version of a product
- Restrictions that apply to a specific release

In a 2002 commentary about *Producing Quality Technical Information*, the handbook on which this book was based, Theo Mandel compares a Web site to a garden. He says, "You must pull weeds and replace plants and flowers on a timely basis or the garden will not attract and interest its viewers." Although Web information is especially prone to becoming outdated, the garden analogy works well for all types of technical information that you develop.

If you take the time to plan and plant a garden, don't you also take the time to water it, fertilize it, and pull weeds from it? Technical information can become outdated and inaccurate just as a garden can wither and become overgrown with weeds due to the lack of attention. Just as a good gardener replaces flowers and plants and pulls weeds, you need to occasionally replace information that was once but is no longer valid with new information.

The following Web page provides general information about the InfoDBase product:

Original

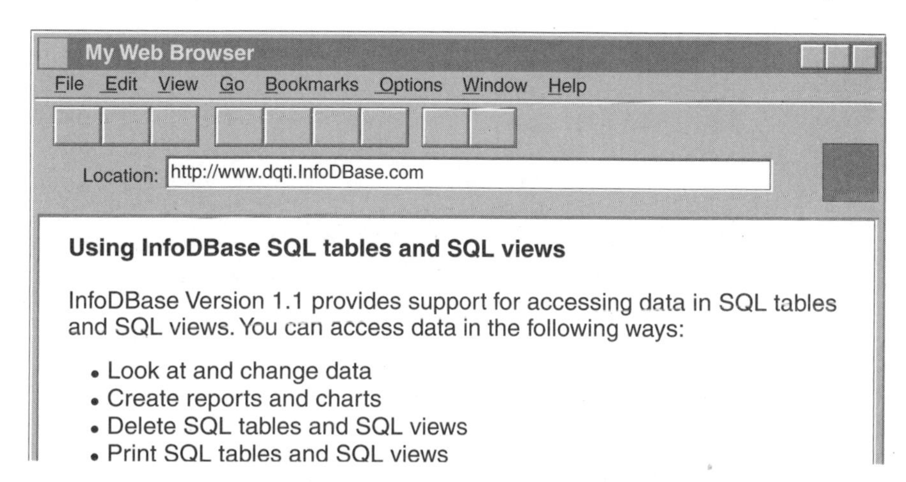

Revision

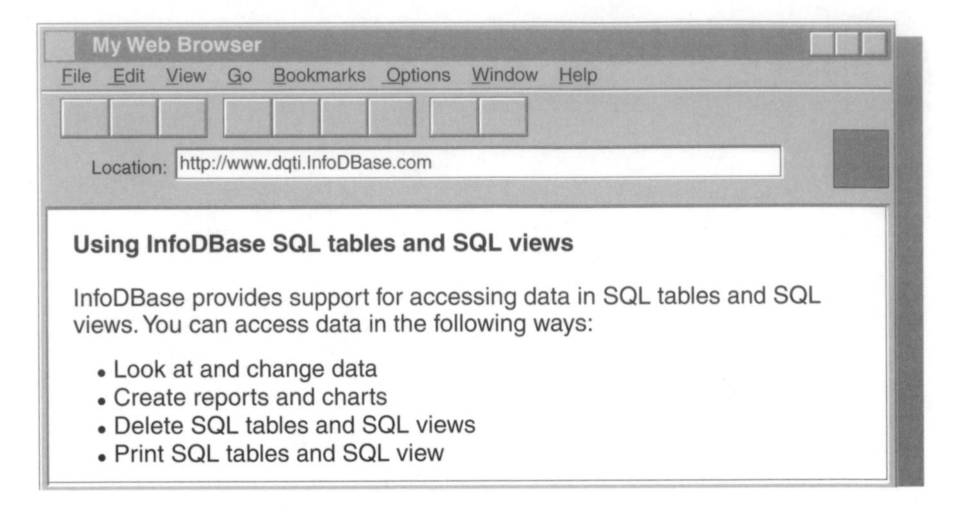

Specifying the release number for a product, as the original Web page does, might be accurate for this release, but the release number could look inaccurate if it's not changed in the next release. If a statement about a product will continue to be true in future releases, write the statement in a general way, as in the revised Web page.

Information can also become outdated during the development cycle for a product, even before it is available. For example, when a specifications document is available, writers need to incorporate information from it. However, this document can get out of sync with the actual product, even if the team records changes in the design. Changes creep in, and people forget to mention them. Ensure that your own information accurately describes the product by checking that the information matches what users will see and experience.

Accuracy problems can result when you include graphics that show all or part of the user interface if the interface changes after the information is finalized. The visual effectiveness guideline "Avoid illustrating what is already visible" on page 280 discourages inclusion of interface elements except when doing so is necessary for users to complete a task.

Assume that a writer determines that the following menu needs to be included in the information and that the writer relies solely on the technical specification:

Original

Play/Pause	Space
Stop	Period
Skip Back	Page Up
Skip Forward	Page Down
Fast Forward	Ctrl+Right
Rewind	Ctrl+Left
Preview	Ctrl+V
Go To. . .	Ctrl+G
Language	▶
Volume	▶

Suppose that the specification that the writer relies on includes the original menu, with the **Fast Forward** and **Rewind** choices in that sequence. The developer, however, decides to change the sequence of the menu items without informing the writing team. The original menu that the writer includes is now inaccurate.

Revision

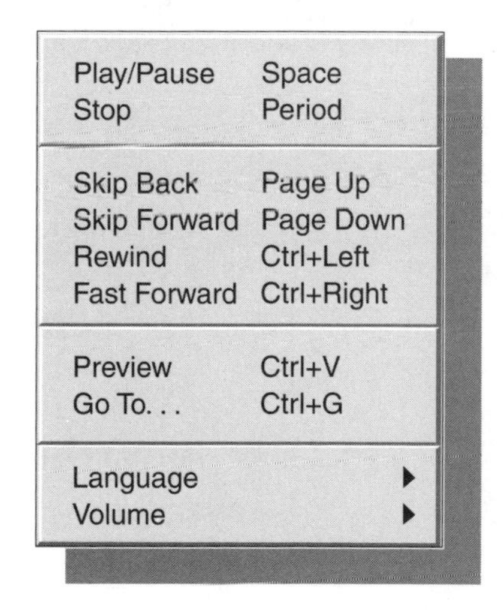

In this case, the revised menu actually matches the interface. A writer who assumes that the specification is correct and includes the original menu in information for users inadvertently introduces an accuracy problem. A writer who checks the actual menu can detect the error and correct the menu in the user information, as shown in the revised menu.

As more and more technical information is delivered on the Web, the challenge to keep that information current is even greater than it is for printed information. Users of the Web expect that the information that they see is current because they know that the Web can be updated at any time. As a result, users are much less inclined to tolerate out-of-date information on the Web than in print.

If you deliver technical information to users on the Web, you can satisfy users' needs for information currency and accuracy by following these tips:

- ❑ Set a regular update schedule and follow it. Identify a qualified backup writer to handle updates when you are unavailable.
- ❑ Indicate the date and time of the last update for the Web page so that users know how current the information is.
- ❑ Post the schedule for updates on the Web page, if possible, so that users know when the information will next be updated.
- ❑ Include a feedback mechanism on the Web page so that users can communicate problems to you. Handle this feedback in a timely manner.

Suppose that a user views the following Web page in the end of March 2003:

Original

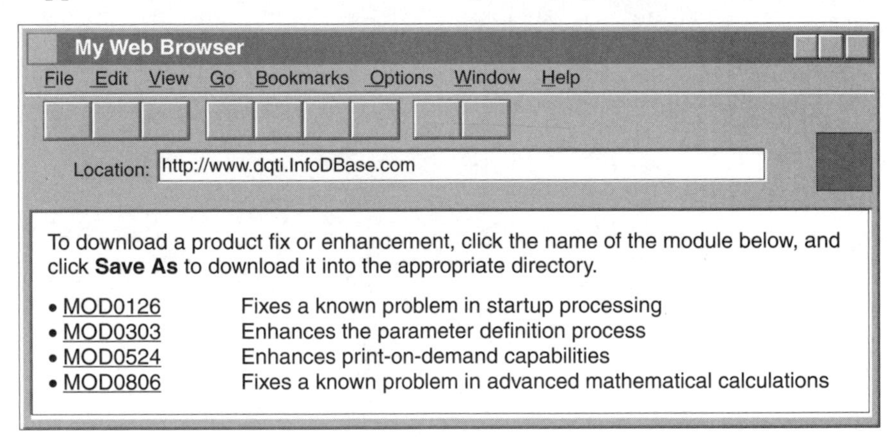

By viewing this original Web page, the user cannot determine whether the information is current.

Revision

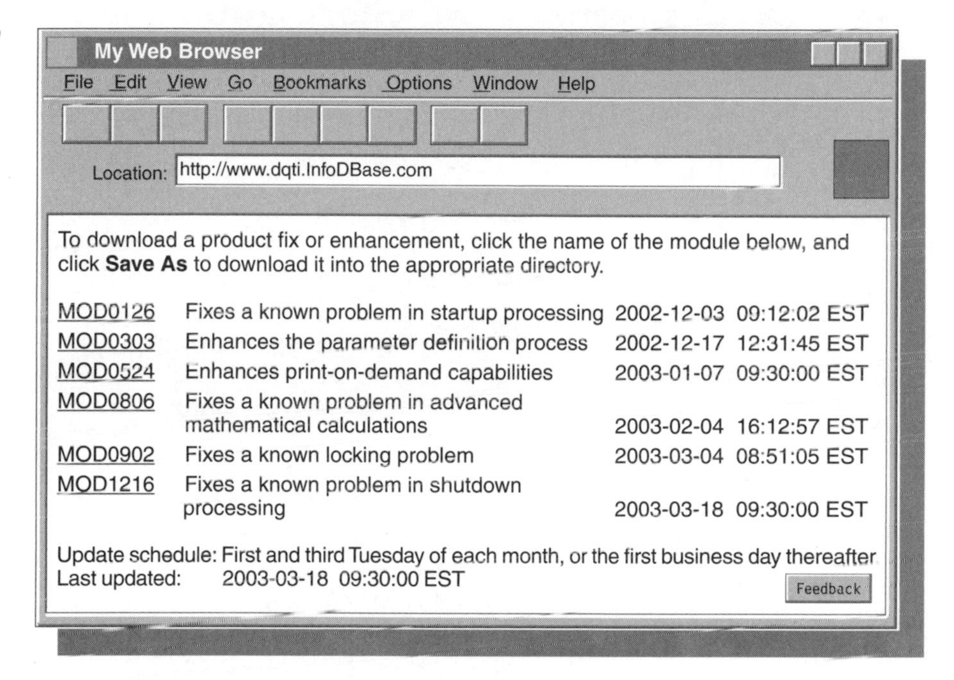

In the original Web page, the list is incomplete, but the user has no way of knowing this. No update schedule is posted, no last-updated date is provided, and no feedback mechanism is available. Users of the revised Web page, however, can trust the currency and accuracy of the information to a much greater degree because of the inclusion of these features.

Regardless of whether information is on the Web or in other formats, technical information that is not current is not accurate. Therefore, include only information that is current, avoid including information that is likely to become quickly outdated, and validate the currency of existing information.

Maintain consistency of all information about a subject

Many accuracy problems occur because the writer updates information in one place but not in other places where it appears. Faced with inconsistent information, the user doesn't know which information is correct.

When you determine that certain information needs to be repeated word for word, you can take advantage of the best way to ensure information consistency: reuse the information.

The simplest form of information reuse involves *copying* the information from one place and pasting it somewhere else. Choose this technique only if the reused information is very short and is unlikely to change in the future.

For longer topics that you do not want to copy verbatim, use a better approach for reuse: single source. *Single source* refers to the use of the same source, with few or no changes, for different outputs. Unless otherwise noted, the word *reuse,* when used in this book, refers to the use of a single source of information, not to the copying of information.

When you use single source, you can design and develop a topic and then use it in multiple contexts. For example, you can use a single topic in more than one medium for information delivery, such as HTML- or XML-based information centers, a Web-based document, online help, and a printed book. Rather than wasting your time synchronizing multiple copies of the same information, you can spend your time ensuring that the topic is of high quality and is accurate and relevant for each information unit that includes it.

Figure 3 on page 59 shows how you might reuse a single topic that provides a list of information resources to users.

Figure 3. Reusing information in multiple contexts.

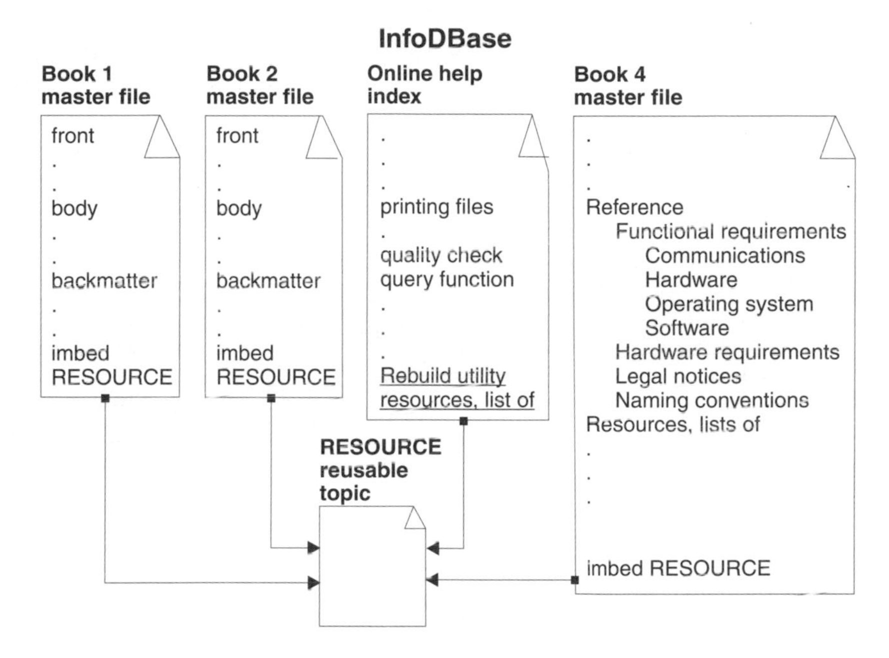

When updates are required, you can make the changes in one place—the source text—rather than updating multiple copies of the text. Whenever a title or referenced Web address changes, the writer of the list of resources updates the source text, once. If the InfoDBase team had used a copy-and-paste approach, multiple InfoDBase writers would need to make the change. This duplicated effort not only increases the amount of maintenance work for the team, but it also increases the chance of inconsistency and inaccuracy.

Regardless of how you approach reuse, accuracy is best served if you do not change the reused information. However, sometimes you need to update some of the text in one of the information units. One advantage of using a single source is that, if your authoring tool supports conditional text, you can include or exclude certain information. By using conditional text, you can specify the conditions for which certain portions of the source text are to be included in or excluded from different information units. Conditional text provides increased flexibility in reusing information for multiple contexts.

The following guidelines can help you maintain consistency:

- ❑ Reuse information when possible.
- ❑ Avoid introducing inconsistencies and eliminate those that you find.

Reuse information when possible

By reusing shared information, you ensure that the user is not confused about what is correct. You also need to change that information in only one place.

Table 1 shows several ways that you can reuse information.

Table 1. Opportunities to reuse information

Opportunity for reuse	Method for reuse
Help text for interface elements that are used in several product windows	Single source for shared help text, accessed by links from individual product windows
Text of messages, used by the product as it displays messages and in printed or online information	Single source for message text, called by the product and by the book formatter
Information on the Web and in printed or other online formats	Single source, formatted separately for Web and print delivery
Online help and related printed information	Single source for technical information topics that are called by a Web browser or by the product for display as help and that are used by the book formatter
Boilerplate text for the front matter of multiple books, which might be delivered in printed, online, or Web formats	Single source for boilerplate text, used by the formatter of multiple books

Not only does using a single source improve the accuracy of information, but it also eliminates the time that is needed to synchronize similar information.

Avoid introducing inconsistencies and eliminate those that you find

Unfortunately, not all related information can be reused. You must proactively synchronize related information that can't be reused.

A good general rule for making information consistent is to minimize the number of places where you include the same information, as in the following examples:

- In a procedure that requires the user to enter the same command more than once, consider giving the full syntax only once (TRACE SYSTEM INFODBASE NEW DEBUG LOG=YES), and then referring to it elsewhere as "the TRACE command shown in step 1." If you repeat the command in each place, you risk inconsistency that might make the user unsure of which format is correct.

- ❑ When you're writing for the Web and want to repeat information from another Web site that you don't control, you can use either of the following approaches:
  - — Paraphrase the content and link to it. The advantage of this approach is that when the information changes (as it is sure to do), you won't need to update it on your Web page, and you won't be responsible for an inconsistency that is perceived as an inaccuracy. The disadvantages are degraded performance for users who need the information fast and the risk that the linked-to Web page will disappear, causing a broken link.
  - — Copy the information to your Web page. The advantage of this approach is improved performance for users who need the information. The disadvantage is that you need to keep your copy of the information in sync with the source.

Whenever information is inconsistent in two or more places, the user's confusion is like your confusion in using a recipe that starts with "Preheat the oven to 325°" and ends with "Bake at 375° for 40 minutes." You don't know which temperature is correct.

Sometimes you need to include technical information in more than one place, often in more than one format. For example, perhaps you include complete sample tables, one of which is shown in Figure 4 on page 62.

Figure 4. Keeping related information synchronized.

Complete Employee table (EMP)

EMPNO	FIRSTNME	LASTNAME	DEPT	HIREDATE	JOB	EDL	SALARY	COMM
000010	CHRISTINE	HAAS	A00	1975-01-01	PRES	18	52750.00	4220.00
000020	MICHAEL	THOMPSON	B01	1987-10-10	MGR	18	41250.00	3300.00
000030	SALLY	KWAN	C01	1975-04-05	MGR	20	38250.00	3060.00
000060	IRVING	STERN	D11	1973-09-14	MGR	16	32250.00	2580.00
000120	SEAN	CONNOR	A00	1990-12-05	SLS	14	29250.00	2340.00
000140	HEATHER	NICHOLLS	C01	1976-12-15	SLS	18	28420.00	2274.00
000200	DAVID	BROWN	D11	1992-03-03	DES	16	27740.00	2217.00
000204	KAREN	RUFFNER	D11	2000-09-15	DES	18	28990.00	2270.00
000220	JENNIFER	LUTZ	D11	1991-08-29	DES	18	29840.00	2387.00
000320	RAMLAL	MEHTA	E21	1985-07-07	FLD	16	19950.00	1596.00
000330	WING	LEE	E21	1976-02-23	FLD	14	25370.00	2030.00
200010	DIAN	HEMMINGER	A00	1970-01-01	SLS	18	28420.00	2274.00
200140	KIM	NATZ	C01	1976-12-15	ANL	18	28420.00	2274.00
200340	ROY	ALONZO	E21	1987-05-05	FLD	16	23840.00	1907.00

You refer to the sample tables in your information where you give examples that rely on these tables. The following passage is an example that shows how to list names of employees in a specific department from the sample Employee table:

Original

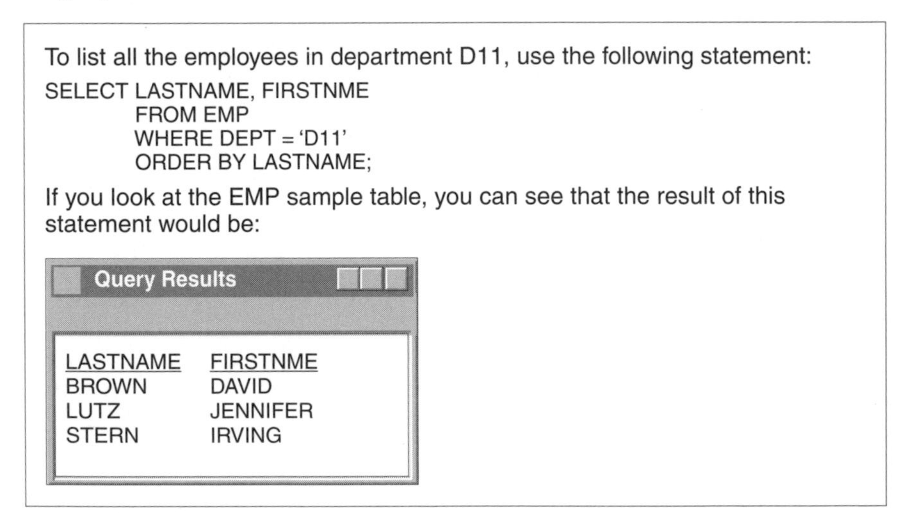
To list all the employees in department D11, use the following statement:

```
SELECT LASTNAME, FIRSTNME
       FROM EMP
       WHERE DEPT = 'D11'
       ORDER BY LASTNAME;
```

If you look at the EMP sample table, you can see that the result of this statement would be:

Users of the original passage who take the time to look at the sample Employee table to anticipate the results of the statement will be confused. They can see from the sample table that department D11 has four employees, but the result table in the original passage shows only three employees in department D11.

Revision

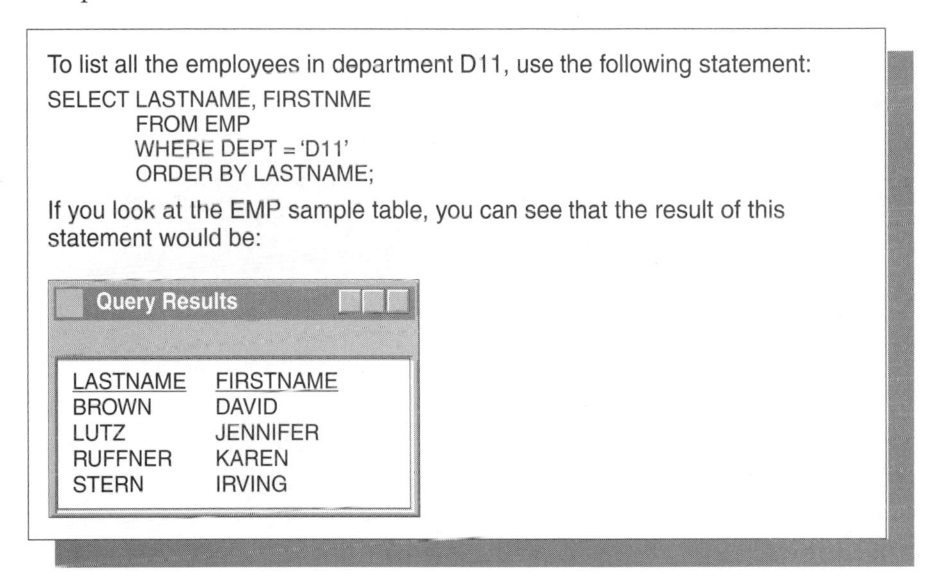

To list all the employees in department D11, use the following statement:

```
SELECT LASTNAME, FIRSTNME
        FROM EMP
        WHERE DEPT = 'D11'
        ORDER BY LASTNAME;
```

If you look at the EMP sample table, you can see that the result of this statement would be:

Query Results

LASTNAME	FIRSTNAME
BROWN	DAVID
LUTZ	JENNIFER
RUFFNER	KAREN
STERN	IRVING

The revised passage accurately lists all four employees in the department; the example is consistent with the sample table and is therefore accurate.

One habit that can help you maintain the consistency of related information is to ask yourself each time that you update information: "Should I make any other changes as a result of this one?" Asking yourself this question might help you think of other parts of your information that need to be changed.

Another good practice is to insert comments into the source for your topics (if your authoring tool allows) or into a separate history file for your project. These comments can identify interdependencies within information that you write, or between your information and the information that your colleagues write. This practice is especially helpful on large projects with many writers and many different but related information units.

For example, consider the preceding situation in which the SELECT statement output was inconsistent with the Employee sample table. In this case, the writer could have avoided the consistency error by seeing an author note in the source for the sample Employee table. The note could indicate that certain examples depend on the content of this table. When changing the sample Employee table, the writer knows to also ensure the consistency of dependent examples in other topics. If you follow this practice, choose a format for your comments that helps you search for them in the future. For example, you might precede each comment with your initials or with a recognizable string of characters.

Other inaccuracies occur when the names or headings of elements are incorrect, such as when the content of a figure or table is inconsistent with its caption or surrounding text. Headings should accurately describe the content of the topic. The column headings of a table should be consistent with and accurately reflect the actual content of the column's cells. Topic titles should accurately reflect the content of the topic.

Suppose that you are a user who is working with the InfoDBase software product, and you aren't sure when in the process you can add columns to an index key. You go to the help and find a help topic like this one:

Original

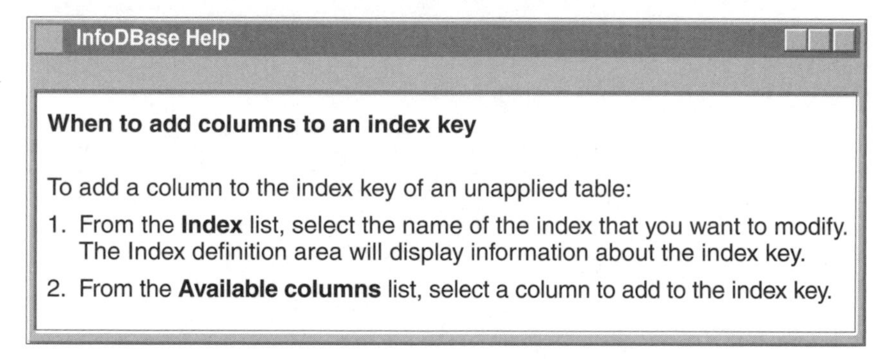

The title of the original help topic implies that the topic explains when you can add columns to an index key. When you see this title in an index of help topics, you probably believe that you've found the information that you're looking for. Instead, it explains how to add the columns.

Revision

InfoDBase Help

When to add columns to an index key

You can add one or more columns to an index key when:

- The database management system (DBMS) is offline.
- The entire system is offline for maintenance reasons.

If you try to add columns to an index key when the database is running, unexpected errors might occur.

The revised help topic actually provides the information that the title promises. The title accurately describes the content of the topic.

Information that is labeled with inaccurate or ineffective headings can also cause retrievability problems, which you will read more about in the retrievability guideline "Provide helpful entry points" on page 257.

When you provide guidelines or rules for users, and you include an example to support those guidelines, make sure that your example adheres to the guidelines, as in the following reference topic for the date command:

Original

date command

Use the date command to set a date for a row of data.

Syntax

```
date mm.dd.yyyy
```

Values for the month and day must each be two characters. A value for the year must be four characters. The values must be separated by periods.

Examples

```
date 5.24.02
```

```
date 08.6.04
```

In the original topic, the syntax rules say that the values for month and day must each be two characters, and that the year value must be four characters. However, the first example shows the month as one character and the year as two characters. The second example shows the day as one character and the year as two characters. This type of inaccuracy can be a major problem because many users focus on examples and don't read the surrounding text.

Revision

date command

Use the date command to set a date for a row of data.

Syntax

```
date mm.dd.yyyy
```

Values for the month and day must each be two characters. A value for the year must be four characters. The values must be separated by periods.

Examples

```
date 05.24.2002
date 08.06.2004
```

The examples in the revised topic correctly show the month, day, and year.

Make sure that related information is consistent. Reuse information by using single source when possible. When reuse is not an option, try not to introduce any inconsistencies when you add information. Eliminate inconsistencies in the existing information that you and your reviewers find.

Use tools that automate checking for accuracy

Although the style chapter addresses typographical and grammar errors, these problems can also affect the accuracy of the information. Many tools are available to help writers identify typographical errors and grammatical errors, and other types of errors such as invalid cross-references.

Typographical and grammatical errors can have a variety of effects. When users notice a minor error, their overall satisfaction with the information might decline slightly. If they notice many errors, they probably question the overall accuracy of the technical information. If the errors are in code examples, user procedures, statements of valid values, or syntax definitions, users are likely to stop reading the information or perhaps stop using the product. As Snyder and Spool mention in "Docs in the Real World," errors in technical information can become "trust-breakers."

The following help topic includes several errors:

Original

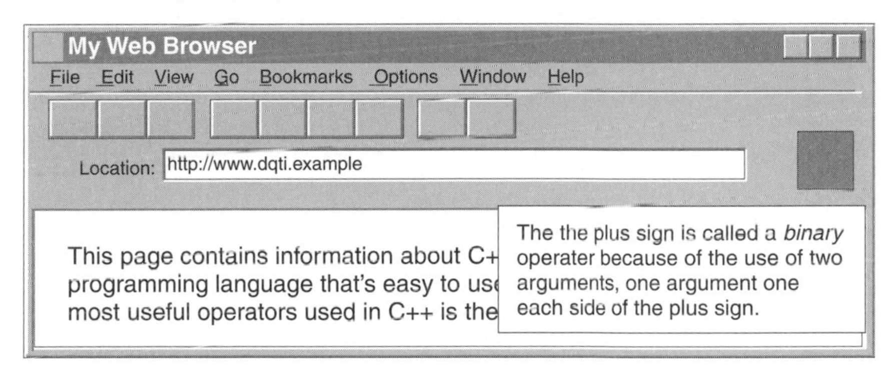

After reading the text of the original help topic, users are much less likely to trust the accuracy of the other information. This help topic demonstrates several problems:

- The word *the* is duplicated.
- The word *operator* is misspelled.
- The word *on* is misspelled as *one*. Most automated tools would not catch this type of error, which reaffirms the need to have information edited by a human editor and not just by automated tools.

Revision

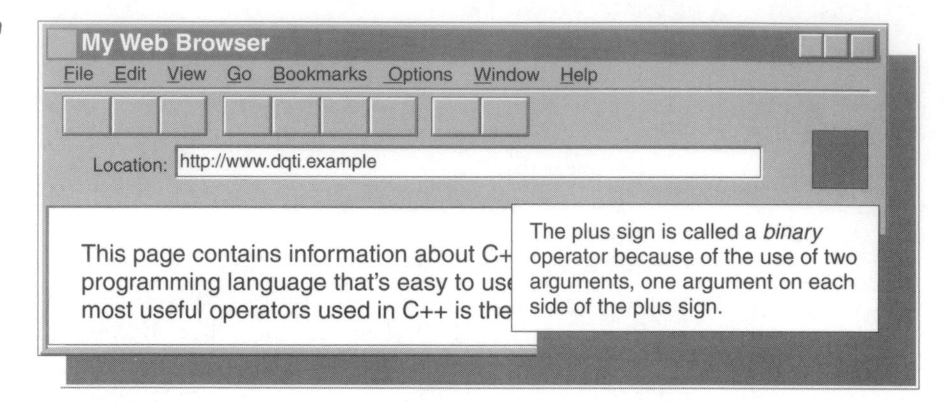

The revised help topic is free of these errors, so users are more likely to trust the accuracy of the information.

If you are like most writers, you have probably made these kinds of mistakes, at least in informal correspondence. However, you can easily find and correct these problems by using automated tools. In the preceding example of the help topic, your spelling checker can find the misspelled word *operater*. You can also use a grammar-checking tool to find other problems. Some tools can check for passive constructions, duplicate words, vague referents, long sentences, and other problems in the information. You might even have access to a single tool that can find all of these problems. Use whatever tools you have to ensure that your information contains no easily detectable errors.

When you use a tool for checking spelling, you can add words to the tool's dictionary. This process is usually as simple as clicking an **Add to dictionary** button. This capability is important to you as a technical writer because your writing almost certainly needs to include terms that are not in the spelling checker's dictionary. When you add words to the dictionary, be careful to do so accurately. Although the process of adding words to the tool's dictionary is usually simple, the task can become tedious and possibly even mindless, especially when the passage that you are checking has many unrecognized words. If you aren't careful, you might inadvertently add a misspelled word to the dictionary, and this action will result in undetected spelling errors.

Check the accuracy of references to related information

You need to pay particular attention to the following kinds of references:

- References to Web locations. To help users find referenced Web information quickly, use the specific Web address for the information. Remember that Web addresses sometimes change. If you know you won't be able to frequently verify that the addresses are valid and update your information accordingly, consider also including a high-level Web address for the company or organization that is responsible for the target information.
- References to printed material. Include the title and the author or publisher.

Internal cross-references. For online information, make sure that each link goes to the correct information. For printed information, make sure that internal cross-references point to the correct information. Publishing software takes care of most cross-reference problems by dynamically rebuilding cross-references, but you still need to check that the referred-to section contains the information that you want the user to find. For more information about using effective links, see the retrievability guideline "Ensure that a link describes the information that it links to" on page 264.

Whenever you update your information, verify all existing references. Sometimes, information that you refer to in a Web page, online document, or printed book might no longer be where you think it is.

Ensuring that target Web pages (Web pages that you cite or link to) are accurate can be especially challenging. A Web page whose address is valid one day might change the next day, through no fault of your own. As information on your Web site ages, the number of broken links from that site tends to increase, as owners of the target sites upgrade servers and do other things that cause their Web addresses to change. Therefore, you are responsible for maintaining and fixing the links and references to target Web pages yourself.

The following Web page contains a broken link:

Original

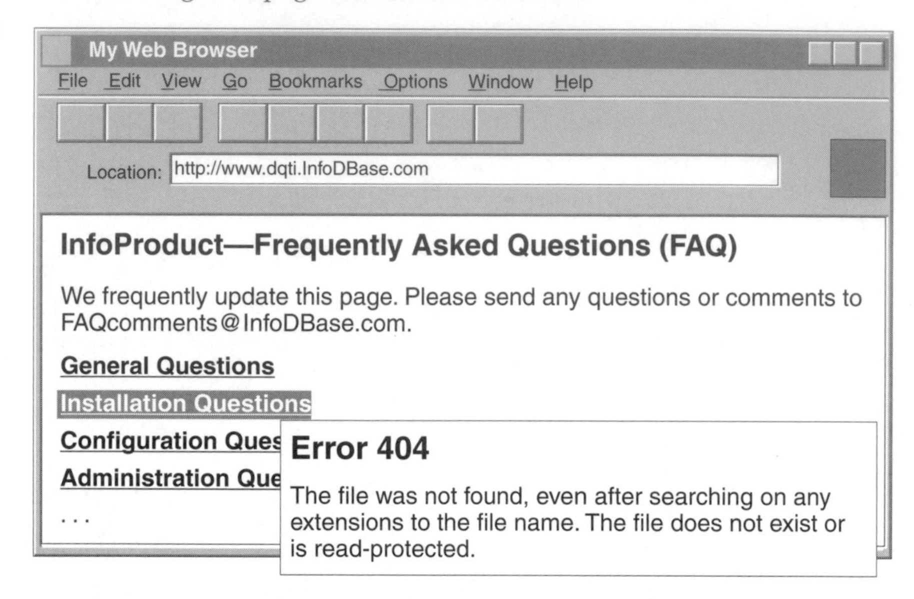

In the original Web page, the link to the Installation Questions Web page fails, and the user receives an error message instead of a Web page.

Revision

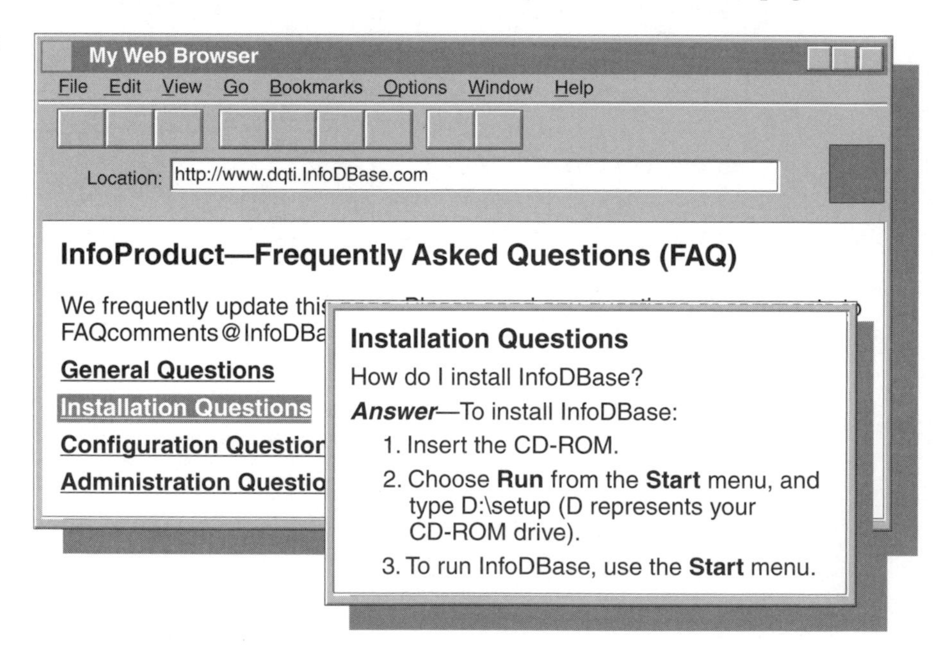

The link in the revised Web page works as designed and takes the user to the intended information.

Using a link-checking tool is a good way to find broken links. If you can find a new address for the target site, fixing the link is easy. Fortunately, many Web site owners now leave behind a trail to the new address (sort of like an automated forwarding address). As a result, even if a target Web address in your information is not valid, users are redirected automatically to the new Web address. Some link-checking tools, possibly the one that you use, can indicate when a link is being redirected to another address. When you learn of a redirected link, you should update the link to the current address for two reasons:

- The automatic forwarding is sometimes only temporary.
- When users know that they are being redirected, they might be concerned about the currency of your information because you haven't updated the link.

Sometimes, though, the target page goes away completely or the specific information is no longer available on the Web. In cases like these, you need to delete the link and update the surrounding text.

Think about how frustrated you are when you take a reference at face value and discover that it does not lead you to the information that you want. To avoid causing that kind of frustration for your users, take the time to verify all references in the information that you develop.

In sum

Although developing accurate technical information is often not easy, the task is vital to users who rely on the information to do their jobs. This chapter provides guidelines to help you write information that is free of technical errors and that is both current and consistent.

After you resolve major accuracy problems with the information, inform users of the improvements to rebuild users' trust in your information.

Use the guidelines in this chapter to ensure that technical information is accurate. See the examples in the chapter for practical applications of these guidelines.

When you review technical information for accuracy, you can use the checklist on page 73 in two ways:

- ❑ As a reminder of what to look for, to ensure a thorough review
- ❑ As an evaluation tool, to determine the quality of the information

You can apply the quality rating in the third column of the checklist to the guideline as a whole. Judging by the number and severity of items you found, rate the information on each guideline for this quality characteristic. You can then add your findings to "Quality checklist" on page 387, which covers all the quality characteristics.

Although the guidelines are intended to cover all the areas for this quality characteristic, you might find additional items to add to the list for a guideline.

Guidelines for accuracy	Items to look for	Quality rating
Write only information that you understand, and then verify it.	• Procedures are correct; they work when a user follows them. • Descriptive information is accurate.	1 2 3 4 5
Keep up with technical changes.	• The final information includes late changes in the product. • Updates of Web information occur at an acceptable frequency. • Trademark lists are current. • Information from previous editions is current and valid. • Interface descriptions are consistent with the actual interface. • References to specific products are current.	1 2 3 4 5
Maintain consistency of all information about a subject.	• Duplicated information is consistent. • No conflicts exist between separate but related topics. For example, sample output matches actual output; graphics and examples are consistent with supporting text; and examples of guidelines and rules are accurate.	1 2 3 4 5
Use tools that automate checking for accuracy.	• Examples, syntax descriptions, procedures, and other types of information are free of spelling errors. • Text is free of grammatical errors.	1 2 3 4 5
Check the accuracy of references to related information.	• Cross-references are correct. • Cited information contains the relevant information. • Titles and order numbers in references are accurate.	1 2 3 4 5

Note: The scale for the quality rating goes from very satisfied (1) to very dissatisfied (5).

Chapter 4

Completeness

A successful book is not made of what is in it, but what is left out of it.
—Mark Twain

From the user's point of view, completeness means that all of the required information is available. A subject is covered completely when all of the relevant information is covered, each subject is covered in sufficient detail, and all promised information is included.

Ensuring completeness involves more than checking items off of a list—it starts with building the right list of what to include. To determine whether the information that you write is complete, you must first know your audience. You need to understand the purpose of the information that you are creating. For example, will users be doing a specific task? learning a concept? making a decision?

Writers have often created technical information that is *too* complete, including everything about a subject. Writing about product function instead of about users' tasks leads to overly complete information. Know your audience. What is their level of expertise? A Web site for computer novices should provide detailed information to help them navigate the site, whereas a Web site for system administrators who need product support requires only a summary of navigation.

One approach to writing technical information is to reduce extraneous or irrelevant information and to provide the user with only the essentials. This approach is called *minimalism*. The minimalist approach results in not only reduced volume of information, but also in information that doesn't get in the user's way. With a minimum of information, the user can independently explore a product after learning some basic concepts and tasks.

Aspects of completeness overlap other quality characteristics, especially task orientation, organization, and retrievability. Knowing your users and what they're trying to accomplish helps you know what information they need and when they need it. Providing information at the wrong time makes it irrelevant.

Completeness also overlaps with visual effectiveness. Writers usually need to use illustrations to describe complex technical concepts. By contrast, using too many illustrations can obscure the necessary information. For example, an installation document that includes screen captures of each installation window is likely to confuse its users with too much detail—detail that they can already see for themselves on their computers. Because of the size of most screen captures, the important information that the users need is probably spread over several pages or screens, instead of fitting on one or two.

Determining and creating the right amount of information is a balancing act. Through task analysis and by using the product, you learn exactly what information users need. As you learn what your users need, you add or subtract information. Doing usability tests can help you identify where there is too much information and where information is missing.

To make information complete, follow these guidelines:

- ❑ **Cover all topics that support users' tasks, and only those topics.**
- ❑ **Cover each topic in just as much detail as users need.**
- ❑ **Use patterns of information to ensure proper coverage.**
- ❑ **Repeat information only when users will benefit from it.**

Cover all topics that support users' tasks, and only those topics

Users of technical information need information about how to use a product to do their job tasks. This guideline is related to the quality characteristic of task orientation, because you cannot decide what topics to cover without identifying your users and evaluating the tasks that those users are to do. By knowing your users and their tasks, you know what topics are relevant.

The minimal task topics are descriptions of the major tasks, their subtasks, the reasons or conditions for doing them, and how to respond to expected errors. Early in your writing project, do a task analysis to determine what tasks and subtasks your users will need to do. Later in the project, you can use the task analysis to evaluate whether your information covers the necessary topics:

- ❑ Have you covered all major user tasks?
- ❑ Have you described all the subtasks that users need to do to complete the major tasks?
- ❑ Have you explained the reasons or conditions that are associated with optional tasks?
- ❑ Have you described what to do if something goes wrong?

If you can answer yes to these questions, your information is more likely to cover all the topics that the user needs.

The following table of contents lacks an important word-processing task, saving a file, probably because the writer failed to thoroughly identify the users' tasks:

Original

Creating a file
Editing text
  Selecting text
  Inserting text
  Overwriting text
  Deleting text
  Highlighting text
  Locating text
  Moving text
  Copying text
Opening an existing file
Printing a file

Revision

Creating a file
Editing text
- Selecting text
- Inserting text
- Overwriting text
- Deleting text
- Highlighting text
- Locating text
- Moving text
- Copying text

Saving a file
Opening an existing file
Printing a file

Although most users of word-processing programs want to save their work, the original table of contents omits the step of saving a file. The revised table of contents includes this important task. This example shows how easy it is for writers who are familiar with a product or task to make assumptions that shortchange the user. Always rely on a thorough task analysis rather than your own assumptions.

Sometimes you need to explain concepts as you introduce a subject, explain a process, or highlight benefits. Do not confuse concepts with information about a product's internals. Concepts help users to learn, because they provide necessary vocabulary or aspects of the big picture. For example, a folder is a concept in a word-processing program, but how that program represents and manipulates a folder internally is not of interest to users.

Even with true conceptual information, you must make sure that the user really needs it. Ask yourself what the user will do with the information. For example, will the information help the user make a decision or do the task at hand? When users install a product, they probably need information about the installation options so that they can decide which ones they need and what values to choose. When they shop on the Web, they need information about the cost and timing of their shipping options so that they can evaluate the overall cost of their purchase and determine whether the purchase will reach its destination in time. Information that isn't relevant to the user's task is probably extraneous.

Cover each topic in just as much detail as users need

Including a topic doesn't necessarily mean that you have covered it adequately. The goal should be to include enough detail for users to do their tasks, but no more detail than they need. While you're analyzing tasks, you might not be able to anticipate all the details of a task; later, when you're writing, questions of detail will arise.

Judgments about the completeness of detail are usually more subjective than judgments about which topics to include. For example, deciding whether to include a topic about troubleshooting is easy, but determining whether to describe error situations that rarely occur is not. Writers who "put themselves in their users' shoes" can usually make good decisions about how much detail is needed.

To understand what information is needed, you must know something about your audience. For example, are your users experienced or inexperienced with your product or with similar products? Knowing the experience level of your audience helps you determine what information is relevant.

Writing about a product is easiest when your audience is homogeneous—either all experienced or all inexperienced. On one hand, if your information is about a new product, you can assume that your audience is made up of novices, at least with this product. On the other hand, if your information is about a well-established product that has been on the market for a long time and sells to relatively few new users, you can assume that most of your audience is experienced.

The challenge that writers frequently face is to successfully write to several different levels of experience. You don't want to bog down the experienced user with lots of introductory information. However, you also don't want to make assumptions that cause major problems for novice users, as in the following topic:

Original

Building InfoTool files

Complete the following steps to build InfoTool files:

1. Log in to the control point host or managed host as root.
2. Change to the directory that contains the driver build script.
3. Run the driver build script.
4. Change to the directory that contains the watchdog build script.
5. Run the watchdog build script.

Revision

Building InfoTool files

Complete the following steps to build InfoTool files and automatically copy the necessary files onto the workstations in your network.

1. Log in to the control point host or managed host as user `root`.

   `login root`

2. Change to the directory that contains the driver build script. For example:

   cd *system_dir/infotool_dir*/drivers/inft64/obj

   In this example, *system_dir/infotool_dir* represents the directory path where you installed InfoTool.

3. Run the driver build script.

   `build_inft_drv`

4. Change to the directory that contains the watchdog build script.

5. Run the watchdog build script.

   `build_inft_wdog`

The writer of the original passage assumed that users knew all the commands and object names that are necessary to do the procedure. If your audience is experienced and does this procedure frequently, then this assumption might be valid. (Even experienced users might require additional detail for tasks that they do infrequently, such as installing a product.) In the revised passage, the writer has included necessary detail for users who don't know the commands and object names but still need instructions.

Figure 5 on page 81 provides a general guideline for the level of detail that you should provide for different types of tasks and audiences.

Figure 5. Novice and experienced users need different detail levels for the same tasks.

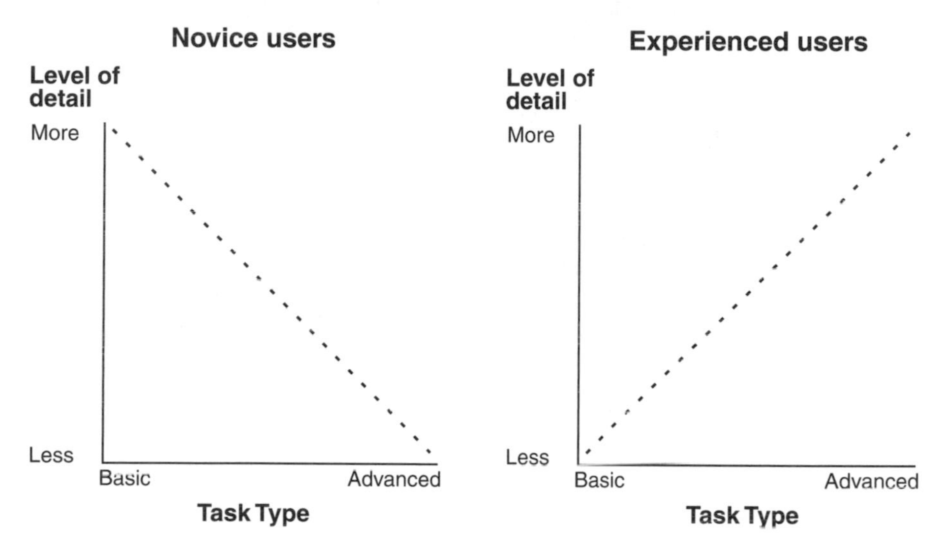

If your audience has a mix of experience levels, you need to decide how to satisfy the varied information needs, without causing problems for any group. These are some approaches that you might select:

- ❑ Create separate concept, task, and reference topics. Link from task topics to the concept and reference topics so that users can get what they need, regardless of their experience level.
- ❑ Write both online and printed information primarily to an experienced audience, but provide extensive online help to assist novices.
- ❑ Put the information into two different formats, such as advanced topics for experienced users and a tutorial for novices.
- ❑ Integrate the information within one online information system or one printed document, but separate the introductory topics from the advanced topics (such as in different windows, chapters, or parts).
- ❑ Integrate the information, but graphically distinguish between information for different experience levels, for example, provide labeled tips or shortcuts for experienced users. Some books separate the technical detail for the experts into boxes, sometimes called *sidebars*, with special graphical icons to attract users' attention.

After you are familiar with your audience, use the following guidelines to cover each topic in just as much detail as users need:

- ❑ Include enough information.
- ❑ Include only necessary information.

Include enough information

When you develop information, or when you're editing information that you or others have written, look for logical holes, areas where some information might be missing. Certain types of information are vulnerable to problems of incompleteness, including:

- Information about possibilities
- Procedures
- Links and cross-references
- Programming syntax

Information about possibilities

If you mention or explain one of many possible situations, mention or explain other likely situations. Otherwise, your user is left wondering about the others. For example, when you describe events with words such as *usually* or *normally*, users might wonder what happens when the unusual or abnormal occurs. Mention not only common occurrences, but also the uncommon ones, especially if users need special information to handle the uncommon situation.

For example, the writer of the revised passage on page 80 could add information about what to do if the build process fails (in steps 3 and 5), such as: "If the build fails, verify that you changed to the correct directory and that the directory contains the necessary files. For a list of necessary files, see the topic, "Files produced during the build process."

Procedures

When you write user instructions, follow every possible path in the procedure. For a complex procedure with many possible paths, you might use an illustration to help orient users. The more decision points in a procedure, the more helpful an illustration like that in Figure 6 is likely to be.

Figure 6. Navigation flowchart for installation procedures.

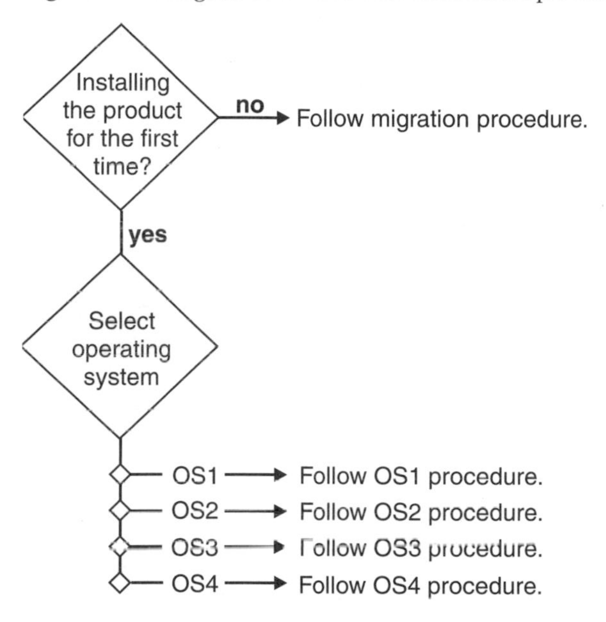

Links and cross-references

When you refer users to other information, either online or printed, ask yourself what users need to know when they decide whether to pursue this information. Don't merely give a title; indicate specifically why you think the source is worth investigating. For example, instead of writing, "For more information, see *InfoDBase User's Guide*," write, "For information about indexing in InfoDBase, see *InfoDBase User's Guide*." Don't expect the user to read the entire referenced source just to figure out what information you thought might be helpful.

When you use a hypertext link in an online document or Web page, and when the text of the link is part of a sentence, make sure to include in the link the words that indicate the type of information to which the link takes the user. For example, if your link takes the user to information about online purchasing of bicycle parts, make the text of the link "purchase bicycle parts," rather than "parts" or "bicycle parts." If you include only nouns and no verb, the shopper will probably assume that the linked-to information defines the term as a list of parts rather than explains how to purchase them.

In printed information, rather than include a heading from another source, use descriptive words that are likely to remain in the index even if the heading changes from one edition to the next. Descriptions of the content of cross-references also help you validate the accuracy of the cross-references when you update your information in future editions.

Programming syntax

When you document the syntax of commands, calls, statements, or macros, ask yourself how much detail the programmer really needs. For example, if you are explaining a graphical technique that you will use for showing command syntax, your explanation should cover every element of that syntax, even the one-character keyword separators.

Regardless of the situation, try to identify and provide the level of detail that your audience needs.

Include only necessary information

Too much information also creates problems of completeness. Unnecessary detail slows down the user, making needed information more difficult to find and use.

The concept that you can create information that is too complete relates directly to the minimalist approach to technical writing. This approach was described by John Carroll in his 1990 book, *The Nurnberg Funnel: Designing Minimalist Instruction for Practical Computer Skill*:

> The key idea in the minimalist approach is to present the smallest possible obstacle to learners' efforts, to accommodate, even to exploit, the learning strategies that cause problems for learners using systematic instructional materials. The goal is to let the learner get more out of the training experience by providing less overt training structure.

Although Carroll originally focused on learners and tutorials, his techniques have been applied more broadly.

Often, too much information confuses users by making it difficult to decide what is important. In the following topic, the writer has included a lengthy list of message identifiers, even though the list doesn't help users with any task:

Original

Reporting an error message problem

Each InfoTool message has an identifier. Message identifiers that begin with ABU indicate programming errors that you can probably correct without help from a service representative. For information about ABU messages and how to correct the errors, see *InfoTool Programming Reference*.

When you receive an error message, follow these steps:

1. Check the third letter of the message identifier. If the letter is U (as in ABU1114E), see step 2. If the letter is S (as in ABS1225E), see step 3.
2. If the third letter is U, you can probably fix the problem yourself. For help with a particular message, see *InfoTool Programming Reference*, which lists all InfoTool messages.
3. If the third letter is S, call 1-800-555-1111 and tell the representative the message identifier. The representative will work with you to resolve the problem quickly. All 150 ABS messages are listed below:

```
ABS1112E    ABS1118E    ABS1132E    ABS1218E    ABS2111E
ABS1113E    ABS1119E    ABS1133E    ABS1219E    ABS2112E
ABS1114E    ABS1120E    ABS1134E    ABS1220E    ABS2113E
ABS1115E    ABS1121E    ABS1135E    ABS1221E    ABS2114E
ABS1116E    ABS1122E    ABS1136E    ABS1222E    ABS2115E
```

and so on

The original topic lists all 150 ABS messages that the InfoTool product issues, even though users can't independently correct the problems that these messages identify. Not only is this a potential accuracy problem (as new messages are added to the product), but it also makes the information longer than it needs to be, and more importantly, does not help users.

The following revised passage excludes the list of messages, which in this case is extraneous, and efficiently tells users what to do if they receive an ABS message:

Revision

Reporting an error message problem

Each InfoTool message has an identifier.

When you receive an error message, follow these steps:

1. Check the third letter of the message identifier. If the letter is U (as in ABU1114E), see step 2. If the letter is S (as in ABS1225E), see step 3.
2. If the third letter is U, you can probably fix the problem yourself. For help with a particular message, see *InfoTool Programming Reference*, which describes all InfoTool messages.
3. If the third letter is S, call 1-800-555-1111 and tell the representative the message identifier. The representative will work with you to resolve the problem quickly.

This pair of topics points out a challenge that writers sometimes face. You might have a difficult time making a complex product seem easy to use. In this case, the writer's job would be much easier if the descriptions of all or most of the product's error messages included actions that the user could take. Forcing users to call a service representative for 150 error situations is far from ideal, but the products for which technical writers provide information aren't always ideal.

Writing about a difficult-to-use product is both a challenge and an opportunity for you as a writer. When you write about a product that is difficult to use, work with the product designers or programmers to improve the usability of the product. Usability testing can be helpful here, as it is in many other writing situations.

You might have noticed an additional change to the revised passage—the introductory paragraph that repeated information that is in the steps was also deleted. Many introductions to topics contain more detail than users need, as in the following topic:

Original

Navigating windows by using the keyboard

Find the Tab key in the upper left area of your keyboard.

Try pressing this key. It moves the cursor between the two fields.

Find the Backspace key in the upper right corner of the alphanumeric area.

Try pressing this key. It erases by going backward from where the cursor is.

You can also erase characters by pressing the long key, called the Spacebar, at the bottom of the keyboard. However, the Spacebar erases by going forward from where the cursor is.

Revision

Navigating windows by using the keyboard

When you enter data in a window, the primary keys that you use are the letter keys, number keys, Tab key, Spacebar, and Backspace key.

Use the Tab key to move from one field in the window to another.

If you make a mistake when entering data, erase it by using the Backspace key or Spacebar.

The original passage includes too many details, both textual and graphical, about how to use a standard keyboard. Unless this is a product that teaches a user how to type, these details are extraneous. The graphical elements break a related visual effectiveness guideline, "Use graphics that are meaningful and appropriate" on page 279, by illustrating what is clearly visible to keyboard users. The revised topic tells users what keys they generally use to enter data in a window, but it excludes unnecessary details.

Too much detail is frequently the result of writing to more than one type of user or to no particular user at all. Sometimes the writer includes extra detail for the secondary audience, which can distract the primary audience.

The terms *primary audience* and *secondary audience* refer to the readers of a specific book or topic; these readers are a specific subset of all of your product users. If you are writing a tutorial, your primary audience consists of novice product users, and your secondary audience consists of more experienced product users who, for example, need to learn about a new function. Typically, you make these distinctions as you analyze your audience and related requirements when you design your information.

Ask yourself: Does this information have a secondary audience? If so, how important is that audience? How (and how much) do they differ from the primary audience? Can the information that they need be omitted or moved elsewhere?

The following passage shows printed installation information for a product that users can install by using a graphical interface or by using a command-driven program. Presumably, most users will install the product by using the graphical interface and associated online help. Therefore, the primary audience of the printed installation information is individuals who use the command-driven installation program.

Original

Configuring an InfoInstaller control point

To configure an InfoInstaller control point, you can use either the graphical program (Infinstg) or the command-driven program (Infinsty). If you use the graphical program, you must have installed InfoFinder.

Preparing to configure a control point

Before you configure an InfoInstaller control point, you need to:

1. Determine (or create) the InfoInstaller control database. If you are using Infinstg, right-click the **CDB** icon to see the name of the control database. If no **CDB** icon is on your desktop, you need to create the database. If you are using Infinsty, go to a command line and enter:

   ```
   create infinsty database
   ```

   If this database already exists, you will receive an error message.
2. *and so on*

In the original passage, the first step addresses both the primary audience and the secondary audience, even though the actions that these users take differ.

The following revised passage shows that two separate procedures have been created. The primary audience is not burdened with information that is directed at the secondary audience, because users in the secondary audience can access the information that they need while they are using the graphical interface.

Revision

> **Configuring an InfoInstaller Control Point**
>
> To configure an InfoInstaller control point, you can use either the graphical program (Infinstg) or the command-driven program (Infinsty). If you use the graphical program, you must have installed InfoFinder.
>
> To use the graphical program, double-click the Infinstg icon on your desktop, and follow the online directions.
>
> If you are using the command-driven program, you must do some preparation before proceeding with the configuration.
>
> **Preparing to configure a control point**
>
> Before you configure an InfoInstaller control point, you need to:
>
> 1. At a command line, enter:
>
>    ```
>    create infinsty database
>    ```
>
> 2. *and so on*

This pair of topics shows an important distinction between the primary audience for certain information and the primary users of certain product function. When information has a specialized audience that is different from most users of the product, ignoring that majority or relegating them to a secondary position can be difficult. The first two paragraphs of the revised topic acknowledge the majority of product users (the secondary audience for this information) and direct them elsewhere. Thereafter, the information targets its primary audience, who use the command-driven program.

Use patterns of information to ensure proper coverage

You can eliminate extraneous information and identify redundancies by analyzing user tasks and by using and testing the information. Another way to determine whether you have covered all of the relevant topics and details is to develop a pattern for structuring your information.

A *pattern* is an organizational device that you can use to make information consistent. Therefore, this guideline is similar to the organization guideline "Organize information consistently" on page 232 and the clarity guideline "Present similar information in a similar way" on page 136.

One pattern that you apply to your information comes from conventions or standards that are based on the policies of an industry, company, or organization or on laws. For example, books usually include some standard parts that frame your information, like covers, copyright and other legal statements, and a table of contents. Web pages often require standard headers and footers.

Standard parts supply only a framework. Patterns within your information help users become comfortable with the organization so that they can focus on the content that they need. A good way to develop a pattern is to start with these questions: What does a user need to fully understand a concept? What does a user need to know to do the task? What does a user need to know to use reference information? From the answers, you can define a pattern for concept, task, and reference topics. For a task topic, the pattern might be:

Heading
Introduction
Prerequisites
Task steps
Links to related tasks

After you decide on a pattern of information, you might choose to create a template. A *template* is a structured type of pattern that is useful for ensuring that all standard parts of parallel, or similar, topics are included. A template is like a cookie cutter, which is good for replicating a standard object in a consistent way. The style guideline "Follow template designs and use boilerplate text" on page 200 describes templates in more detail.

Contextual help topics or Web pages for similar topics should contain the same types of information at the same level of detail. If contextual help is available on some editable fields, it should be available on all editable fields. If a help topic includes a summary list of tasks with links to more information, similar summaries should also have links. Consider the following pair of help topics:

Original

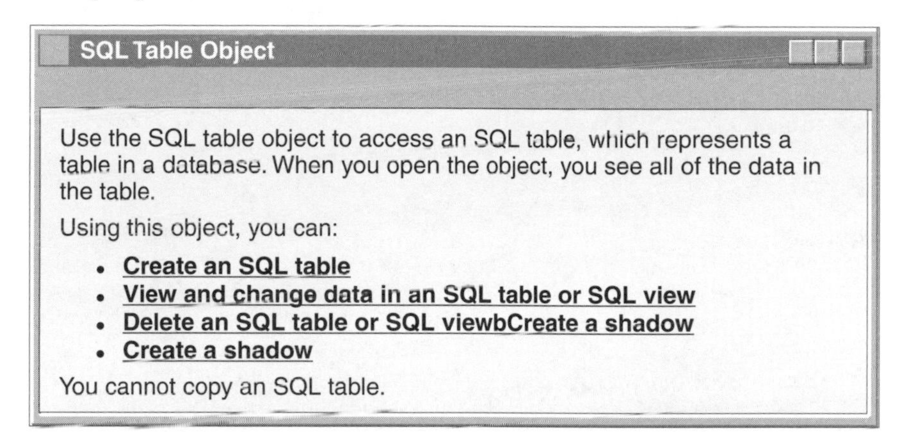

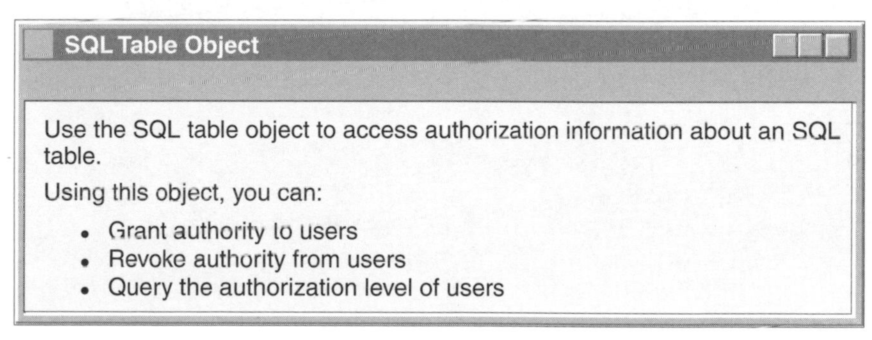

In the original pair of help topics, the first topic uses links, and the second does not. The following revised pair of help topics demonstrates the pattern by using links for a task summary:

Revision

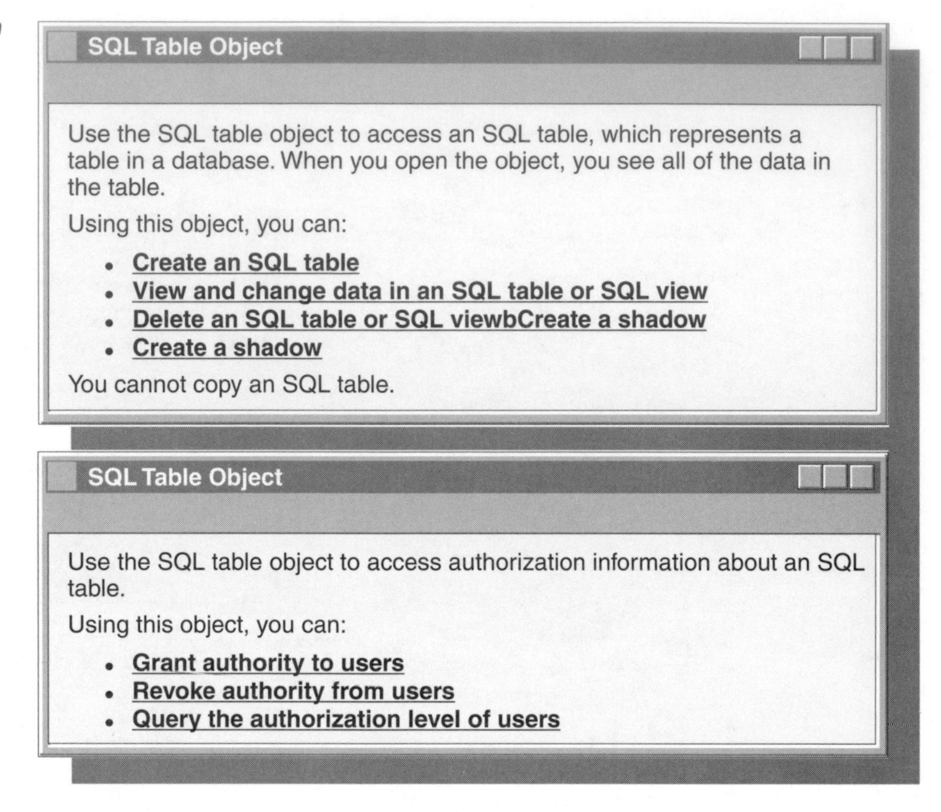

Patterns are also useful for explaining how to do a task. Task information that is standard from one product or environment to another, such as installation, might have many of the same subtasks in common.

In the following pair of topics, assume that InfoDBase and InfoTool are companion products that run in the same operating environment and are installed in the same way:

Original

Installing InfoDBase

To install InfoDBase, do the following tasks:

1. Review hardware and software prerequisites.
2. Install the InfoDBase program files.
3. Configure the InfoDBase environment.
4. Verify the installation and configuration.

****

Installing InfoTool

To install InfoTool, do the following tasks:

1. Review hardware and software prerequisites.
2. Identify prerequisites that aren't installed and install them.
3. Install the InfoTool program files.
4. Configure the InfoTool environment.
5. Verify the installation and configuration.

In the original topics, the overviews of the installation task for InfoDBase and InfoTool differ in that the first overview omits the step of installing prerequisite products before installing the program.

Revision

Installing InfoDBase

To install InfoDBase, do the following tasks:

1. Review hardware and software prerequisites.
2. Identify prerequisites that aren't installed and install them.
3. Install the InfoDBase program files.
4. Configure the InfoDBase environment.
5. Verify the installation and configuration.

****

Installing InfoTool

To install InfoTool, do the following tasks:

1. Review hardware and software prerequisites.
2. Identify prerequisites that aren't installed and install them.
3. Install the InfoTool program files.
4. Configure the InfoTool environment.
5. Verify the installation and configuration.

The revised topics follow a pattern that ensures that both sections contain the same kind of information, in the same sequence, and at the same level of detail. Users who install both programs can be confident that all the relevant information is present.

When you use patterns to ensure completeness, however, make sure that you use a correct pattern. If the writer of the previous help topics had used the InfoDBase structure as the pattern for the InfoTool topic, the necessary information about installing prerequisite products would be missing from the installation instructions for both InfoDBase and InfoTool. Before you use a given information pattern for other information, verify that the pattern is valid by collecting user feedback or doing usability testing.

As in task topics, patterns in reference topics are very useful. The following partial table of contents is for command reference information:

Original

Statements
ALTER
  Syntax
  Examples
  Usage notes
BACKUP
  Syntax
CREATE
  Syntax
  Examples
  Usage notes

Revision

Statements
ALTER
  Syntax
  Examples
  Usage notes
BACKUP
  Syntax
  Examples
  Usage notes
CREATE
  Syntax
  Examples
  Usage notes

In the original table of contents, the section on the BACKUP statement omits some important information—examples and usage notes. The revised table of contents includes these pieces of information. Using a pattern or template for reference information can help you ensure that you cover all similar topics at the same level of detail.

When you promise information to your users, your information is incomplete until you fulfill that promise. You might overtly promise information, for example, by stating, "This section describes all of the configurations that you can choose for the InfoDBase product." Patterns implicitly promise information to users by creating expectations.

Repeat information only when users will benefit from it

Repetition can be good or bad. Repetition is good when it appropriately emphasizes and reinforces important points or enables users to avoid an unnecessary branch to another topic or page. Repetition is also good when you don't know where users might enter your information, as is the case on the Web. Needless repetition is bad; extraneous details waste reading time and might give a wrong impression about what is important.

The following pair of topics shows the value of repetition. The original topic contains no repetition, but provides links to relevant information.

Original

To select a different set of fonts, follow these steps:

1. Change your directory to *fonts*.
2. Review the list of available font files and decide which one best meets your needs.
3. Copy the font file that you want to use to your printer.

Related tasks:

- ❑ Changing directories
- ❑ Copying files

Related reference:

- ❑ Available fonts

Revision

To select a different set of fonts, follow these steps:

1. Change your directory to *fonts*.

   ```
   cd fonts
   ```

2. Review the following list of available font files and decide which one best meets your needs:

   Arial
   Brittanica
   Courier
   Engravers
   Futura
   Hogbold
   Times New Roman

3. Copy the font file that you want to use to your printer.

   ```
   copy filename.fnt lpt1
   ```

You wouldn't want to copy everything from another information source. However, as the retrievability guideline "Minimize the effort that is needed to reach related information" on page 265 indicates, do not send the user to different topics for information about a simple instruction or for a short list of available fonts. The revised passage includes the necessary details in one place and doesn't force the user to go to other topics.

When you repeat information unnecessarily, you not only interrupt and delay the user's task, but you might also confuse a user and waste space—especially online, where space is limited. The following help topic shows how unnecessary details can obscure a task:

Original

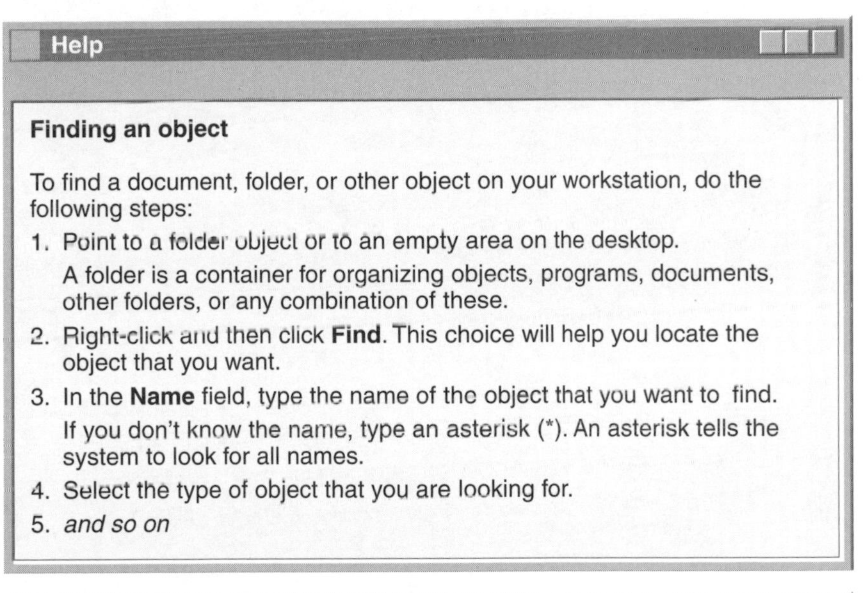

Revision

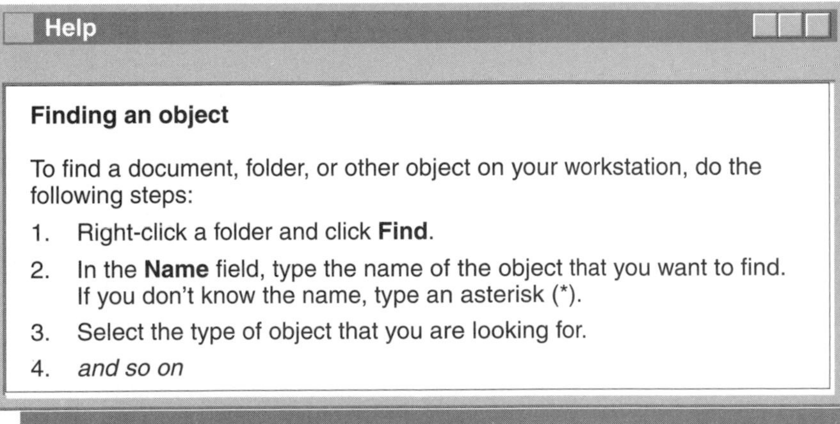

In the original help topic, steps 1 through 4 repeat details that are covered by other contextual help topics, making this procedure longer than necessary (possibly forcing the user to scroll). The revised help topic provides only the appropriate information for this task.

When you know your audience and the tasks that they need to perform, you can decide when to repeat information to facilitate completing a task.

In sum

Information is complete when it includes all of the topics that users need to successfully complete their tasks. These topics include neither too much nor too little information, and they repeat information as necessary. You can use patterns in your information to help identify missing information.

This chapter includes guidelines to help you ensure that technical information is complete but does not provide more information than users need. Refer to the examples in the chapter for practical applications of these guidelines.

When you review technical information for completeness, you can use the checklist on page 100 in two ways:

- ❑ As a reminder of what to look for, to ensure a thorough review
- ❑ As an evaluation tool, to determine the quality of the information

You can apply the quality rating in the third column of the checklist to the guideline as a whole. Based on the number and severity of items you found, decide how the information rates on each guideline for this quality characteristic. You can then add your findings to "Quality checklist" on page 387, which covers all the quality characteristics.

Although the guidelines are intended to cover all areas for this quality characteristic, you might find additional items to add to the list for a guideline.

Guidelines for completeness	Items to look for	Quality rating
Cover all topics that support users' tasks, and only those topics.	• Information contains no extraneous topics such as product internals. • Topics cover the information that users need for doing tasks. • The amount of conceptual information is appropriate for the audience. • Information that doesn't directly support a user task includes an explanation for why that information is necessary.	1 2 3 4 5
Cover each topic in just as much detail as users need.	• Explanations contain enough detail. • Procedures or tasks are complete; no steps are missing. • Syntax rules are complete. • Syntax descriptions include all elements. • Primary topics provide more detail than secondary topics. • Only the most common uses are described in detail. • New concepts are sufficiently developed. • References to related information are meaningful. • Introductory information is sufficiently detailed for the audience.	1 2 3 4 5
Use patterns of information to ensure proper coverage.	• Explanations of parallel information provide a similar level of detail. • Standard parts are available.	1 2 3 4 5
Repeat information only when users will benefit from it.	• Information is repeated when it is important for the user. • Information does not include unnecessarily repeated sections.	1 2 3 4 5

Note: The scale for the quality rating goes from very satisfied (1) to very dissatisfied (5).

Part 2

Easy to understand

Whether technical information is easy to understand depends mainly on how it is presented at the level of small structures such as words and sentences. It can also involve larger structures such as examples and scenarios.

Chapter 5

Clarity

Easy reading is damn hard writing. —*Nathaniel Hawthorne*

Clear information is information that users can understand the first time. They don't need to reread it to untangle grammatical connections, sort out excess words, decipher ambiguities, figure out relationships, or interpret the meaning. Clarity in technical information is like a clean window through which you can clearly see the subject.

The first time that most writers try to write about something, words don't tumble out clearly. Often, to break through a block, writers write as if they were speaking, but this approach introduces problems such as unnecessary words, vague referents, and rambling sentences. People can get visual and auditory cues when you're speaking, but written words are on their own to convey a message.

Clear information is mainly the result of rewriting—replacing, adding, and deleting parts to achieve clarity. Clear information requires close attention to elements such as words, phrases, sentences, lists, and tables to make sure that each participates appropriately in the message. This attention is probably more helpful when you are rewriting than when you are first writing.

Clarity provides the rationale for many decisions about style and for some decisions about visual effectiveness. You might choose, for example, boldface for certain uses of words and italics for certain other uses; consistently applying these style decisions enhances clarity by making information easier for users to understand the first time.

Guidelines for achieving clarity in technical writing have traditionally been based on experience and on studies with native speakers of English, particularly American English. However, because of the explosion of technical information on the Web and the increased use of technical information throughout the world, technical writers must be sensitive also to the needs of an international audience for clear English.

Aspects of clarity also overlap aspects of completeness (especially relevance and too much information) and organization (especially consistency and subordination). Clear information requires a strong focus on what users need to know and when they need to know it. Extraneous information muddies the message. Relationships among pieces of information (whether in sentences, lists, tables, or some other element) must be clear.

To make information clear, follow these guidelines:

- ❑ **Focus on the meaning.**
- ❑ **Avoid ambiguity.**
- ❑ **Keep elements short.**
- ❑ **Write cohesively.**
- ❑ **Present similar information in a similar way.**
- ❑ **Use technical terms only if they are necessary and appropriate.**
- ❑ **Define each term that is new to the intended audience.**

Focus on the meaning

Clear information requires that you focus on what you want to say. What's the point? What do you want the users to do or to know?

When most people write, they need to warm up to a subject. They need to write a while before the words flow and they see what they need to say. A first pass can produce long sentences, imprecise words, unnecessary modifiers, and rambling paragraphs.

Consider how you would focus on the meaning in the following paragraph, which gives an overview of a product for storing and retrieving image data:

Original

> A major company maintains a large personnel database that it basically makes use of for many kinds of employee-related applications such as payroll and benefits. The company now has some plans to extend the database significantly to include current photographs of employees and use the photographs as the basis for a very modern security system. The system is specifically designed to protect secure areas of the company's building from access by any people who really do not have authorization to be there.

Each sentence has over 25 words, but are they all needed? Does each word contribute to the meaning? In an inflated economy people must spend lots of money for everyday items such as groceries or a tank of gas; in this "inflated" passage users must read many more words than the message warrants.

Contributors to the word inflation in this paragraph are:

Imprecise words

"now has plans to," "makes use of," "kinds of"

Intensifying words

"basically," "significantly," "very," "specifically," "really," "some," "any"

When you revise the passage to remove these, you get:

First revision

> A major company maintains a large personnel database that it uses for many employee-related applications such as payroll and benefits. The company plans to extend the database to include current photographs of employees and use the photographs as the basis for a modern security system. The system is designed to protect secure areas of the company's building from access by people who do not have authorization to be there.

This revision is clearer, but the passage is still not as clear as it could be. Many of the nouns are modified, but only a few modifiers are needed:

- "Major" and "large" do not supply useful information in this situation; in fact, they detract from the meaning by limiting the example unnecessarily. Unfortunately, the emphasis in the first part of the sentence falls on these modifiers, as you will find if you read the sentence aloud.
- You don't need "many" as a modifier of the applications if you also have the examples of applications.
- "Current" is not needed because in this context you can take for granted that the photographs should be up to date.
- "Modern" introduces a judgment that is probably not needed in a paragraph that is not intended for marketing the product.
- "Secure" as a modifier of "areas" is confusing because the areas won't be secure unless they are protected.
- "Company's" is superfluous because the context should make clear that the building is the one associated with the company.
- The modifier of "access" is a long phrase that you can shorten just to "unauthorized."

When you remove the unnecessary modifiers, you get:

Second revision

> A company maintains a personnel database that it uses for employee-related applications such as payroll and benefits. The company plans to extend the database to include photographs of employees and use the photographs as the basis for a security system. The system is designed to protect certain areas of the building from unauthorized access.

By focusing imprecise words and removing intensifying words and unnecessary modifiers, you reduce the length of the paragraph by around a third. Now the meaning is clearer, but you can make it even clearer by removing extraneous information. In this context users care that the company is extending the database for image data but not that the company maintains the database or uses it for many employee-related applications. The first sentence eases into the subject, but it is not needed for clarity.

When you remove the extraneous information, you get:

Third revision

> A company plans to extend its personnel database to include employee photographs. The company wants to use the photographs in a security system that will protect certain areas of the building from unauthorized access.

The revised paragraph focuses on the company and the relationship between the personnel database, employee photographs, and the security system.

The revised paragraph has only two sentences. You might find that you have more that you want to say in this paragraph. Perhaps, you could mention the product as a way of storing and retrieving the photographs in the database.

Some writers believe that a good paragraph should have at least three sentences—one to introduce the subject, one to develop the subject, and one to summarize the additional information and lead into a new subject. Such a formula for a paragraph can help you think through the information that's appropriate for a paragraph. However, the formula should not result in extra sentences that get in the way of the message. Clarity is better served by a short paragraph than by one with extra sentences.

Short paragraphs help users scan and read information online, including on the Web. People are likely to skip long or difficult paragraphs, and so they might miss information that they need.

When you focus on what you want to say, you can fix many clarity problems, especially:

- Imprecise words
- Overuse of intensifying words
- Long sentences
- Unnecessary modifiers
- Rambling paragraphs

Make every word count. Eliminate words that do not contribute to the meaning.

As you can see in the following list, an imprecise verb depends on a word or two after the verb to give it meaning. For example, in the combination "have plans," the verb *have* suggests nothing about planning, and the sense of planning has moved to the object of the verb. Information is clearer when a verb conveys the action.

Imprecise verbs	Precise verbs
be in agreement	agree
be capable of	can
carry out an inspection of	inspect
conduct an investigation of	investigate
do a verification of	verify
draw a conclusion	infer, conclude
give an answer	answer
give rise to	cause
have a requirement	require
have knowledge of	know
have plans	plan
have the capability to	can
hold a meeting	meet
keep track of	track
make a distinction	distinguish
make a proposal	propose
make a suggestion	suggest
make changes to	change
make contact with	meet
perform the printing	print
provide assistance	help
reach a decision	decide
render inoperative	break
serve to define	define
show improvement	improve

In addition to making some verbs more precise, you might be able to eliminate some words, such as:

absolutely	completely	particularly	simply
actually	definitely	perfectly	some
any	fairly	quite	specifically
basically	just	really	totally
certainly	of course	significantly	very

Such words are meant to intensify the meaning, but you'll probably find that the meaning is clearer without them. In speaking, people tend to use these words liberally, but in writing, these words quickly lose their effect. When you consider using an intensifying word, try the sentence in your mind with and without the intensifying word; think about the effect on the meaning of the sentence and surrounding sentences.

When you rewrite a sentence or a paragraph, consider how each word contributes to what you want to say. Question whether all the words in a long sentence are needed, and even whether the sentence is needed.

Avoid ambiguity

This guideline could be called "Have mercy on translators and nonnative speakers." These people are more likely than native speakers to have difficulty when information is ambiguous. Writing for an international audience is especially necessary for material that people access on the Web. English is the native language of less than half of Internet users, although around 85% of Web pages are in English. A third of Internet users primarily speak a European language, and around a quarter primarily speak an Asian language.

To avoid ambiguity in what you write, follow these guidelines:

- ❑ Use words with a clear meaning.
- ❑ Avoid vague referents.
- ❑ Place modifiers appropriately.
- ❑ Avoid long strings of nouns.
- ❑ Write positively.
- ❑ Make the syntax of sentences clear.

Use words with a clear meaning

Words that have more than one meaning can be confusing when more than one meaning fits the context. Some ambiguous words are small but important in conveying information about relationships between clauses.

Problems arise when words such as *since*, *once*, *as*, and *while* could indicate either time or something else, such as cause or contrast. Consider the use of *since* in the following sentence:

Original

> Since you created the table, you have authority to change the table.

Revision

> Because you created the table, you have authority to change the table.

A native speaker would probably recognize that *since* in the original sentence applies to cause rather than time. However, using the unambiguous word (as in the revision) is safer than relying on the context to help users make the correct choice.

Because of the richness of the English language, the words that yo
will probably have more than one meaning. However, you can red
confusion over which meaning a word has in a particular context by using the word consistently with only one of its meanings. For example, use "replace the part" to mean only "put the part back" and not also "substitute a new part for the one that you have."

The following list shows some words that are likely to be ambiguous and some words to use instead:

Ambiguous	Clear
as	because
as long as	if
in spite of	regardless of, despite
may	can, might
once	after, when
on the other hand	however, by contrast
since	because
through	finished
while	although, whereas

May and *can* are confusing when they are used interchangeably in technical information. As a child, you probably learned to use *may* when asking for permission. In technical information, *may* rarely has the meaning of permission but is sometimes used ambiguously to indicate possibility or capability, as in the following sentence:

> You may restrict authority to issue certain commands.

Does this sentence mean that the user can restrict the use of the commands or that the user might do so from time to time? A user cannot know, unless the writer has used *may* and *can* conscientiously in the surrounding information and thus prepared the user for the intended meaning here.

Do not use *may* and *can* interchangeably. Reserve *may* for the possibility of doing something and *can* for the ability to do it. In technical information, capability is usually more appropriate to write about than possibility.

Might indicates a greater degree of uncertainty about a possibility and so indicates possibility less ambiguously than *may* does. You could choose to use *might* rather than *may* in many sentences that deal with possibility.

A word that can be used as more than one part of speech can be confusing when the word is used in two different ways close together or when it is used with other such words. For example, each of the following words can be used as a verb or as a noun:

change	document	process	team
design	function	record	work

Consider the following sentences:

Original

Record the date of the record.
The timer function functions to stop certain actions.

The first sentence puts the same word at opposite ends of the sentence, first as the verb and then as the noun. The second sentence puts the noun and the verb next to each other.

Revision

Register the date of the record.
The timer function serves to stop certain actions.
The time function stops certain actions.

Each revision keeps the noun but changes the verb. Two possible verb changes are shown for the second sentence. In general, the use of a word as a verb offers more opportunities for synonyms and paraphrases than does the use of a word as a noun. For example, if the verb *record* should have a less formal sense than *register*, you could use *write* or *remember* instead. Each of these examples is one sentence, but you need to watch for possible confusion in such words also when they are in the same paragraph or in the same topic.

In the following sentence, the nouns are hard to distinguish from the verb:

Original

The team documents design changes.

Revision

The members of the team document the changes in the design.

The revision inserts articles before the nouns (*changes* and *design*) and expands the noun string (*design changes*), making the verb (*document*) more obvious. The change in the subject to *members of the team* also helps reduce the possible interpretation that *team* modifies *document*.

Use words that have clear meanings. Look for other ways that words might be misunderstood, particularly by nonnative speakers of English.

Avoid vague referents

Another clarity problem for translators and nonnative speakers is vague referents. Sometimes when writers use a pronoun to refer to a noun, the noun is vague for any of these reasons:

- It is far from the pronoun, perhaps in another sentence.
- It is among other nouns that might fit as the one being referred to.
- It is not stated explicitly.

Consider the vague referent in the following passage:

Original

> InfoGrammar is a virtual database that you can create from the target file system. It consists of a set of statements that provide input to the projection utility.

Revision

> InfoGrammar is a virtual database that you can create from the target file system. This database consists of a set of statements that provide input to the projection utility.

The original passage is confusing because you're not sure whether *it* refers to InfoGrammar or to the target file system. Is the referent the subject of the preceding sentence or the noun that immediately precedes the pronoun? The revision specifies *database*, clarifying the referent without repeating it.

Consider the referents for the pronouns *each* and *one* in the following sentence:

Original

> You can have multiple catalogs for a single source; however, each can access only one.

Revision

> You can have multiple catalogs for a single source; however, each catalog can access only one source.

In the original sentence, the referents for *each* and *one* are not clear. Native speakers might guess correctly what these words refer to, aided by the order of the nouns in the first clause. However, even native speakers waste time considering whether the clause might mean that "each source can access only one catalog."

A pronoun such as *it*, *this*, or *that* at the start of a sentence can cause confusion, as in the following passage:

Original

> Copy this file so that you can edit and save your version locally. This is especially important when InfoProduct is installed on a network and you want to run a customized version of the product.

Revision

> Copy this file so that you can edit and save your version locally. Having your own copy is especially important when InfoProduct is installed on a network and you want to run a customized version of the product.

The original passage refers to a noun that is only implied in the preceding sentence. The revised passage replaces the pronoun with a noun phrase ("having your own copy").

Check that the antecedents for the pronouns that you use are clear. Also, the style guideline "Use correct grammar" on page 183 discusses some pronoun errors to be aware of as you write or edit.

Place modifiers appropriately

Modifiers include prepositional phrases and relative clauses, as well as adjectives and adverbs. Modifiers that are placed inappropriately cause confusion, as in the following sentence:

Original

> Click the configuration server to which you want to link the list manager servers in the Configure Server Name field.

The phrase "in the Configure Server Name field" is out of place. [illegible]
not where the linking occurs. Instead, that field has the names of [illegible]
uration servers, and so you might revise this sentence as:

First revision

> Click the name of the configuration server in the Configuration Server Name field to which you want to link the list manager servers.

Now you have another problem of a misplaced modifier. The *that* clause should modify "configuration server" and not the field, which the restrictive clause is next to. You have two choices, as in the following revision:

Second revision

> Click the name (in the Configuration Server Name field) of the configuration server to which you want to link the list manager servers.
>
> In the Configuration Server Name field, click the name of the configuration server to which you want to link the list manager servers.

Depending on the context, you could choose to put the intervening phrase in parentheses, or to give the position first and then specify the action.

As another example of a misplaced modifier, consider the use of *consistently* in the following sentence:

Original

> Programs that produce errors consistently require regular maintenance.

The use of *consistently* here could be interpreted in any of the following ways:

- How regularly the errors are produced
- What kinds of errors are produced
- How unvarying the relationship is between programs that produce errors and the requirement for regular maintenance

The revision offers three possible reconstructions of this sentence:

Revision

> Programs that consistently produce errors require regular maintenance.
>
> Programs that produce consistent errors require regular maintenance.
>
> If a program produces errors, it almost always requires regular maintenance.

Users must guess about what the original sentence means. Writers need to foresee this ambiguity and rewrite an ambiguous passage to ensure that the words convey the meaning that was intended. A misplaced modifier like *consistently* is called a *squinting modifier* because it seems to look both forward and backward at the same time.

The word *only* is a big source of ambiguity in technical information because it can modify almost any part of speech and so can occur in almost any position in a sentence. The position of *only* affects the meaning of a sentence, as in the following variations:

> Only an administrator can open the XML files in the latest browser.
> An administrator can only open the XML files in the latest browser.
> An administrator can open only the XML files in the latest browser
> An administrator can open the XML files only in the latest browser.

For clarity, put *only* immediately before the word or phrase that it modifies.

Many writers tend to put *only* near the verb, as in the following sentence:

Original

> The grid is only active when an object is placed or dragged on a form.

Revision

> The grid is active only when an object is placed or dragged on a form.

The revision moves *only* to the position where it emphasizes the conditions under which the grid is active rather than the state of the grid.

Make sure that modifiers are in the correct position relative to what they need to modify.

Avoid long strings of nouns

The English language allows the use of nouns as adjectives. This use helps English supply new terms, but it can also contribute to ambiguity when several nouns are combined.

The names of software or hardware features and functions seem particularly prone to couplings of nouns, such as "work load," "table space," "error message," "access key," and "procedure call." Over time some pairs become accepted as one word, as in "database" and "workstation."

Strings of three nouns are common as names of features and functions such as "program request handler," "attribute properties window," and "network terminal option." These combinations can cause confusion about what modifies what. For example, is the handler for program requests, or is the request handler for programs? Does the middle noun go with the first noun or the last noun, or are the three equal?

If the first pair of nouns in a threesome serves as an adjective for the third noun, consider using a hyphen, as in "character-set identifier." If, however, the combination is well established as having no hyphen, you need to observe that convention.

Do not use a noun string to modify another noun string. The buildup of nouns can quickly become confusing, as in the following sentence:

Original

> Use the input message destination transaction code as shown in the example.

Revision

> Use the transaction code for the destination of an input message as shown in the example.

In the original sentence, determining whether *destination* goes with the first pair or the second pair of nouns is difficult. To make sense of a long string of nouns, you need to insert prepositions and articles.

Be careful when you use an adjective with a string of nouns. In some languages, the ending of an adjective must agree in number and gender with the ending of the noun that it modifies. In English, placement of an adjective (rather than its ending) before a noun usually indicates that it modifies the noun. When users see an adjective that precedes a noun string, they sometimes have difficulty deciding whether the adjective modifies one of the nouns or the set of nouns. Consider the following sentence:

Original

> The sample dynamic plan selection routine is in the Samples folder.

First revision

> The sample routine for dynamic plan selection is in the Samples folder.

The revised sentence sorts out what modifies what by placing the first adjective with the last noun. However, the noun pair "plan selection" is a poor twosome because it masks a verb, as shown in the following revision:

Second revision

> The sample routine for dynamically selecting a plan is in the Samples folder.

Avoid creating noun strings with nouns that should be verbs. "Performance control" and "error recovery" are other examples of such nouns; in these cases, you could use phrases such as "to control the performance" and "to recover from errors."

Consider rewriting strings that have more than two nouns. When you change a long noun string to make the words easier to understand, start with the noun or pair of nouns at the end of the string.

You might be stuck with some noun strings because of existing terminology in a field or in your product. However, be careful not to create new ones, especially ones where a verb should be used instead.

Write positively

Putting more than one negative word in a sentence can make the sentence difficult to understand. If you put two negative words (such as "not unlike") in a clause, you make the clause positive. Research shows that readers have more trouble understanding a sentence that has double negatives than an equivalent positive sentence.

Consider the use of *not* in the following sentence:

Original

> Do not install InfoConnect until you check that your computer does not have these conflicting programs.

First revision

> Before you install InfoConnect, check that your computer does not have these conflicting programs.

The original sentence is difficult because of the negatives in two clauses. The revision changes the negative "not...until" in the main clause to a positive form, making the negative in the dependent clause easier to understand. Telling users what to do is much stronger than only implying it by saying what not to do.

Some negatives mask a need to supply more information. The first revision might be fine for users who don't find any conflicting programs, but what about those who do? Consider the following revision:

Second revision

> Before you install InfoConnect, check your computer for these conflicting programs.

This revision gets rid of "does not have" and sets the stage for information about what to do if you find the conflicting programs on your computer.

Negative words can take any of several forms. The more obvious ones are *no, not, none, never*, and *nothing*. Common combinations with *not* are:

Negative expressions	Positive expressions
not many	few
not the same	different
not different	similar
not unlike	like
not exclude	include
not . . . until	only when
not . . . unless	only if

Another common way for negatives to get into a sentence is through a negative prefix, which is attached to the beginning of a word:

Negative prefix	Examples of words that use the prefix
de-, dis-	detach, deactivate, deemphasize, disentangle, disable, disagree
ex-	exclude, extinct, extinguish
in-, ir-	ineffective, inefficient, ineligible, irregular, irresponsible
non-	nonexistent, nonrestricted, nonsecure
re-	reduce, refuse, reject, reverse
un-	unavailable, undo, unlike, unpredictable, unrelated

Some other negative words that are less obvious have negative meanings, although their form looks positive: *avoid, limit, wrong, fail, doubt*.

You can usually make sentences easier to understand if you replace *not* combinations with one word (even a negative word), as in the following list:

Negative expressions	Preferable expressions
does not have	lacks
does not allow	prevents
does not accept	rejects
not able	unable
not possible	impossible

If you can state a sentence in positive terms, do so. Consider how you would make the following sentence positive:

Original

> Transitions cannot occur between states that are inherited from different classes or between local and inherited states.

Revision

> Transitions can occur only between states that are inherited from the same class or between states that are defined in the object.

The revision gives the same information but avoids the negative construction. This positive approach is particularly important for a sentence that starts a paragraph.

However, some negatives are necessary and useful. For example, the most straightforward way to state a restriction or warning is: "Do not . . . ." Properly used negatives can be effective.

Make the syntax of sentences clear

Some words (such as conjunctions, articles, prepositions, and relative pronouns) carry little meaning in themselves, but they glue together the more significant parts of a sentence. These so-called *syntactic cues* help users, particularly translators and nonnative speakers, understand how the words in a sentence relate to each other. For example, an article (*the, a, an*) can help users recognize a noun, which might be the subject or an object, depending on its relationship to the verb in the sentence. Machine-translation systems also depend on such cues to analyze information.

To make sentences shorter or more informal, some writers drop syntactic cues that they regard as unnecessary, as in the following sentence:

Original

> Each illustration accurately depicts an object, concept, or function.

Revision

> Each illustration accurately depicts an object, a concept, or a function.

Using an article with each noun in a series is often considered optional, particularly when the same article would be used with each noun. In the revised sentence, however, the article changes because one noun starts with a vowel and the next two nouns start with a consonant. Adding the correct articles might make the sentence easier for some users to understand but might sound strange to other users. You need to consider your audience and the situation rather than automatically leave the articles out or put them in.

Many syntactic cues are optional depending on the context. Writers tend to leave these little words out of instructions, lists, glossary definitions, and descriptions in reference information. Consider the following definition:

Original

> **Code point.** Unique bit pattern that is defined in code page.

Revision

> **Code point.** A unique bit pattern that is defined in a code page.

Some writers might readily insert the second *a* but hesitate about adding the first *a*. However, both are needed to clarify that *the* is not the article that has been dropped in either place. The article *the* would suggest that only one unique bit pattern is defined in a particular code page.

The following verbs require the syntactic cue *that* when these verbs take a noun clause as the object:

announce	explain	require	understand
assume	indicate	specify	validate
be sure	mean	suggest	verify
ensure	recommend	suppose	write

Consider the following sentence:

Original

> Ensure each illustration accurately depicts a function.

Revision

> Ensure that each illustration accurately depicts a function.

In the original sentence, users tend to read *each illustration* as a direct object of *ensure* until they reach the verb *depicts*, and then they recognize that the object is really a noun clause. With the addition of *that*, the revised sentence makes clear, immediately after the verb, that a clause is coming.

Another case of possible dropped words involves the verbs *give* and *assign*, which take an indirect object, but writers often leave out the *to*, as in the following sentence:

Original

> Give each illustration a caption.

First revision

> Give a caption to each illustration.

If you didn't know the meaning of *illustration* and *caption* and if you didn't know that indirect objects usually precede direct objects, you might have trouble understanding the original sentence. Although the revised sentence is clear, it's somewhat stilted. You might consider a further revision:

Second revision

> Write a caption for each illustration.

This revision uses a more specific verb than *give* and so avoids the problem.

Perhaps the most troublesome deletion, even for native speakers of English, is *that* when it introduces a relative clause. Many writers have learned to leave it out, as in the following sentence:

Original

> Each illustration accurately depicts the function it was designed to illustrate.

Revision

> Each illustration accurately depicts the function that it was designed to illustrate.

The addition of *that* in the revised sentence helps clarify how the parts relate to each other without sounding stilted.

Consider what happens when the sentence becomes more complex:

Original

> Ensure that each illustration accurately depicts the object, concept, or function it was designed to illustrate, and does so as simply as possible.

First revision

> Ensure that each illustration accurately depicts the object, concept, or function that it was designed to illustrate, and that it does so as simply as possible.

In this case you might be tempted to leave out the second or third *that* if the sentence sounds stilted. Instead, try rewriting the sentence, as in the following revision:

Second revision

> Ensure that each illustration accurately and simply depicts the object, concept, or function that it was designed to illustrate.

A telegraphic style (using only major words to communicate, as in a telegram) can jeopardize readability for many users. In technical information, particularly if it is to be translated or used by nonnative speakers of English, avoid the telegraphic style.

Keep elements short

Wordiness wastes reading time, space, and paper, and it sometimes buries the message. Writing concisely is especially necessary for information on the Web, where reading is 25 % slower than for printed material.

As you saw in the section "Focus on the meaning" on page 105, extra words can get in the way, making users sort through the words to weed out the excess. This section presents more ways to get rid of extra words. To keep elements short, follow these guidelines:

- ❑ Remove roundabout expressions and needless repetition.
- ❑ Choose direct words.
- ❑ Keep lists short.

Remove roundabout expressions and needless repetition

Roundabout expressions use several words where one or two or none will do. In speaking, people often use extra words to gain time to think about the next thing to say. In writing, however, these extra words are like noise and hamper the meaning. How many roundabout expressions can you find in the following sentence?

Original

Given the fact that you have created an object, it can operate in a manner independent of other objects based on the same class.

Revision

After you create an object, it can operate independently of other objects of the same class.

The original sentence uses three roundabout expressions: "given the fact that," "in a manner," and "based on." Such expressions tend to occur when you want to connect important parts of a sentence, as if a simple preposition or conjunction were not enough.

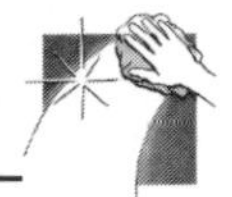

Be suspicious when several words clump together, and consider whether one word might do, as in the following list:

Roundabout expressions	Concise terms
at present	now
at this (that) point in time	now (then)
a variety of	various (*maybe* different, many, several, *or nothing*)
due to the fact that	because
during the course of	during
for the most part	usually
for the purpose of	for
given the condition that	if
in an efficient manner	efficiently
in case of a	in a
in conjunction with	with
in order for	for
in order to	to
in the event that	if (*or* when)
of an unusual nature	unusual
on account of the fact that	because
with regard to	about

Using repetitive words such as *subject matter* is a habit in English that started when people wanted to show that they knew how to say something in more than one language. They would combine an Anglo-Saxon word such as *new* with the Latinate word such as *innovation*.

People still carry on this habit of needless repetition, as in the following sentence:

Original

> You can display a visible grid on forms to help you place components.

Revision

> You can display a grid on forms to help you place components.

If you're displaying the grid, presumably it is visible. The revision removes the needless repetition.

The following list shows words that are often used together, especially in speaking, but their meaning is so close that one word is enough. Cut away the excess.

Needless repetition	Succinct expressions
adequate enough	enough (*or* adequate)
by means of	by
create a new	create a
end result	end (*or* result)
entirely complete	complete
every single	every (*or* single)
exactly the same	the same
group together	group
integral part	part
involved and complex	complex
is currently	is
new innovation	innovation
one and only one	one
period of time	period (*or* time)
plan in advance, advance planning	plan, planning
refer back	refer
repeat again	repeat
share in common	share
sequential steps	steps
subject matter	subject
summary conclusion	summary (*or* conclusion)

Choose direct words

In addition to choosing between using more words and using fewer words, you have a choice when two words have much the same meaning.

Sometimes the English language has two words that have the same meaning but one is shorter. Often the longer word derives from Latin (such as *assist*) and the shorter word derives from Anglo-Saxon (such as *help*). In these cases, the Anglo-Saxon word is usually more direct, as in the following list:

Words derived from Latin	Words derived from Anglo-Saxon
accomplish	do
additional	more
construct, fabricate	make
discover	find, realize
enumerate	list
initiate, commence	begin, start
locate	find
majority	most
perform	do
possess	have
present	give, show, tell
prior to	before
require	need
retain	keep
terminate	end
transmit	send
utilize, employ	use
via	by, through

Whatever the origins of a word, the shorter word is usually more direct, as in the following sentence:

Original

> Modifications to a shared property apply universally to objects on the object list.

Revision

> Changes to a shared property apply to all objects on the object list.

As shown in the style guideline "Write with the appropriate tone" on page 193, your choice of words affects the tone of what you write. Latin-derived words contribute to a formal tone in writing.

You might sometimes decide to use a Latin-derived word. As with intensifying words, such a word will be more effective because of its rarity. Also, you need to be sensitive to the language background of your audience. For example, research into the preferences of French speakers for English vocabulary shows that they prefer Latin-derived English words to words derived from Anglo-Saxon (Thrush, 2001). This finding is perhaps not surprising because Latinate English words are likely to resemble French words.

You might be tempted to replace some Latin-derived words with phrasal verbs, which consist of an Anglo-Saxon verb plus a preposition or two. Here are some examples:

Words derived from Latin	Phrasal verbs
accelerate	speed up
align	line up
complete	fill in, fill out
defer	put off
disconnect	cut off
exclude	leave out
review	look at

Phrasal verbs pose a problem for nonnative speakers of English because many such verbs are not defined in the dictionary or, if they are defined, the dictionary might not include the meaning that was intended. For example, the dictionary recognizes the meaning for *fill out* as "to put on flesh" rather than "to complete." Also, a person can't necessarily determine the meaning of the phrase by looking up the separate words. If you are writing information that will be translated or that will be used by nonnative speakers of English, avoid phrasal verbs that might be confusing.

You need to choose direct words as appropriate for your audience.

Keep lists short

Long lists, especially long lists of tasks or subtasks, can overwhelm users. Users tend to feel more optimistic about succeeding at a task if the steps or subtasks do not run on and on. Users need to comprehend the task as a whole and see how the parts contribute to it.

At a minimum, when users are reading the steps in a task or subtask, they should be able to do the steps without constantly referring back to the information; they should be able to remember what they have read long enough to use it in the immediate situation. Ideally, users who consult a task topic once should be able to remember the steps for a longer time and not need to consult the same topic later if they need to do the task again.

Short-term memory (also known as *immediate memory* or *working memory*) is thought to be a type of memory that people use to store information temporarily, before they either commit the information to long-term memory or forget the information. Research on short-term memory that George Miller reported in 1956 indicated that people can remember seven items plus or minus two. Research since then indicates that short-term memory has less capacity, maybe four to six items (according to Broadbent in 1975), four items (according to MacGregor in 1987), or three items (according to LeCompte in 1999).

Do these changing findings over the decades indicate improved research on a changeless human capacity for short-term memory? Another explanation might relate to a decreasing emphasis on memorization in our schools, where students get fewer assignments (like memorizing all the capitals or the multiplication tables) that involve transferring items from short-term memory to long-term memory. Yet another explanation might be that people are relying more on mechanical memories such as calculators and personal digital assistants for both types of memory.

Whether the downward trend in the results from the research on short-term memory shows improved research, atrophying of a mental ability, greater reliance on mechanical memories, or some other phenomenon, writers of technical information must help users remember even a few items at a time, particularly for short-term use. The better that users can understand the steps the first time, the more easily they should accomplish the transfer to long-term memory if they want to remember the information.

The traditional guidelines on the number of items in a list that users need to remember are:

- Seven items maximum for online information
- Nine items maximum for printed information

Perhaps the safest approach when writing such lists is to limit the total number of items by grouping items. The following passage could benefit from this grouping approach:

Original

To check the accessibility of a document.

1. Install the latest version of InfoReader.
2. In InfoReader, select **Edit —> Preferences —> Accessibility**.
3. On the Content Delivery page, select the **Deliver data in pages** option, and then the number **2**.
4. Select **Edit —> Preferences —> Options**.
5. Clear the **Display Document in Browser** option.
6. Open the document.
7. Listen to InfoReader.
8. If InfoReader does not start reading, press Insert+down arrow.
9. Have InfoReader read the table of contents. To go to a particular page, press Ctrl+n and then enter the number of the page that you want InfoReader to read.
10. If a page contains only text, you can scan the page from paragraph to paragraph by pressing Ctrl+down arrow.
11. Press Ctrl+PgUp or Ctrl+PgDn to move from page to page.
12. Display the Links List and verify that all of the links are listed.

    The pop-up menu closes after you select the command.

This list of 12 items is too long, and some items are confusing because they focus on manipulating the interface without a clear reason. The procedure involves several parts of the product and some prerequisite tasks. Grouping the steps around more general tasks would clarify the procedure. The task orientation guideline "Group steps for usability" on page 38 also supports this approach.

Revision

To prepare to check the accessibility of a document:

1. Install the latest version of InfoReader.
2. Open InfoReader.
3. If you want InfoReader to run faster for large documents, do the following steps to deliver large documents in small increments:
   a. Click **Edit —> Preferences —> Accessibility**.
   b. On the Content Delivery page, select the **Deliver data in pages** option.
   c. Select the number **2**.
4. Click **Edit —> Preferences —> Options**. Clear the **Display Document in Browser** option, so that the document will display in InfoReader.

To check the accessibility of a document:

1. Open the document and listen to InfoReader.
2. If InfoReader does not start reading, press Insert+down arrow.
3. Have InfoReader read the table of contents. To go to a particular page:
   a. Press Ctrl+n.
   b. Enter the number of the page that you want InfoReader to read.
4. If a page contains only text, you can scan the page from paragraph to paragraph by pressing Ctrl+down arrow.
5. Press Ctrl+PgUp or Ctrl+PgDn to move from page to page.
6. Display the Links List and verify that all of the links are listed.

With the grouping of steps by more general tasks, the number of steps has decreased. In the revision, the first set of steps is clearly for preparation, and the second set of steps is clearly the heart of the procedure. Another possibility would be to move the information about keys to a table in a reference topic and link to that information from this more general procedure. Such an arrangement would be particularly useful if several procedures rely on the same information about keys. If this is the only procedure that needs that information, these steps should specify the keys.

Long lists of reference information (often presented in alphabetical order) can be acceptable because users look up information in such lists rather than read them from beginning to end. When users need to remember the items in a list for long-term use, be particularly careful to keep lists short.

Write cohesively

Elements such as sentences, lists, tables, figures, and examples that are clear on their own do not necessarily result in clear information overall. These elements must work together to emphasize the most important ideas that you want to convey. The ideas that are expressed in each element should not have equal weight. Instead, some ideas should reinforce other ideas, and thus contribute to the overall impact of the information.

Make explicit the logical flow from idea to idea. This flow can take many forms, such as from sentence to sentence, from paragraph to paragraph, from a paragraph to a list, table, figure, or example, and back again. At the start of sentences, you can use transitional words or phrases (such as "for example," "therefore," "however," and "as a result"). To integrate lists, tables, figures, or examples, you can introduce them with a sentence and provide meaningful captions for figures and tables, as in the following topic:

Original

Using a grid to place objects

A grid is a set of horizontal and vertical lines that appear on InfoProduct forms. A grid is active only when objects are placed or moved on forms.

You can display a grid on forms to help you place components. You can control the size and visibility of the grid lines. The grid is useful when you draw or place components as they are created or when you move and resize them.

If you want the grid lines to be visible, set the **Visible** property to True. To change the space between grid lines, type new values in the **Height** and **Width** fields. These values measure the distance in twips (twentieths of an inch point) between the grid lines.

Property	Possible values	Default	Effect
Active	True False	True	Makes an object align with the grid lines Lets you place an object anywhere on the form
Width Height	45 to 1485 twips	5 pixels	Width sets the vertical space between lines Height sets the horizontal space between lines
Visible	True False	True	Makes the grid visible Makes the grid invisible

Taken individually, each sentence is clear. However, the paragraphs are not clear, and the table is not integrated. Judging by the first sentence in each paragraph, you can surmise that the first paragraph describes a grid, the second paragraph tells what you can do with a grid, and the third paragraph tells how to use the properties. Looking more closely at the sentences, however, you find that the flow of ideas is:

- Paragraph 1
  - Definition of grid
  - Restriction on when the grid is available
- Paragraph 2
  - Usefulness of a grid
  - Aspects of a grid that you can customize
  - Usefulness of a grid
- Paragraph 3
  - How to make grid lines visible
  - How to change the space between grid lines
  - Measurement of the space between grid lines
- Table
  - Meaning of values for the Active property
  - Meaning of values for the Size property
  - Meaning of values for the Visible property

As you can see, some ideas are repeated: the usefulness of a grid and some property information. However, the passage does not deal clearly with what it means for the grid to be visible and what it means for the grid to be active.

The following revision gradually provides the details while avoiding repetition:

Revision

Using a grid to place objects

You can display a grid (a set of horizontal and vertical lines) on forms to help you place objects. You can control these aspects of the grid:

- ❑ Whether objects automatically align with the grid
- ❑ Whether the grid is visible
- ❑ The size of the vertical and horizontal space between grid lines

You will probably want the grid to be active and visible, so that you can use it as you create objects or move them. However, objects will also align with the grid when it is active and invisible.

Table 1 gives the possible values for the Active, Visible, Width, and Height properties of a grid.

Table 1. Properties of a grid and how to use them

Property	Possible values	Default	Use to ...
Active	True False	True	Make an object align with the grid lines Place an object anywhere on the form
Visible	True False	True	Make the grid visible Make the grid invisible
Width Height	45 to 1485 twips*	5 pixels	Set the vertical space between lines. Set the horizontal space between lines.

* A twip is a unit of measure that represents one-twentieth of a point. There are 72 points in an inch, 1440 twips in an inch, and 567 twips in a centimeter. Use twips when you need a measurement that is independent of the screen resolution.

The flow of ideas in the revision is:

- ❑ Paragraph 1
  - — Usefulness of a grid (including definition)
  - — List of the meaning of the properties
- ❑ Paragraph 2
  - — Relationship of active and visible
  - — Relationship of active and invisible
- ❑ Paragraph 3
  - — Introduction to the table

- ❑ Table
  - — How to make grid lines active
  - — How to make grid lines visible
  - — How to change the space between grid lines

The revision includes needed information without being redundant. It includes the definition of a grid but emphasizes the value of a grid to the user. Online information could link to a definition rather than include it here. The revision also includes a definition of twip but as a note to the table rather than as text; the definition is important only in the context of the information in the table.

The revision relates the text to the table by using a list to emphasize the meaning of the properties. The list treats all three properties, and then a paragraph goes into more detail about the relationship between the Active and Visible properties. This detail is warranted because the relationship is not obvious.

Writers sometimes use transitions less in reference information, which users tend to sample rather than read consecutively, or in online information, where space is at a premium. In these situations, headings are a clear way to change a subject or to express the purpose of a chunk of information.

Present similar information in a similar way

One way to help users understand information the first time that they read it is to present similar information in a similar way. This guideline applies to both the content and the format of technical information. It therefore overlaps the following quality characteristics and guidelines:

- Completeness ("Use patterns of information to ensure proper coverage" on page 90)
- Style ("Create and follow style guidelines" on page 203)
- Organization ("Organize information consistently" on page 232)
- Visual effectiveness ("Use visual elements logically and consistently" on page 295)

Here are some ways to present similar information:

- Use the same notation for the syntax of a programming language in all online and printed information.
- If you enclose some figures in boxes, enclose all the figures.
- If you describe one of the four benefits of a product in terms of a problem that the product solves, describe the other three benefits that way.
- If you explain some procedures through annotated examples, explain them all that way.

You might accomplish the first two items by following style guidelines or using visual elements consistently. The last two items require more attention to the content. You need to decide whether clarity is best served by such consistent treatment of the information.

To present similar information in a similar way, follow these guidelines:

- Use lists appropriately.
- Segment information into tables.

Use lists appropriately

Suit the type of list to the information. Use an ordered list for information where the sequence or priority is important. Instructions in procedures are the most common use of ordered lists in technical information. Use an unordered list for items of similar importance, such as a set of alternatives or parts. Both kinds of lists are used appropriately in the following topic:

Entering the shortcut text

You can add the encoded shortcut text in either of two ways:

- ❑ Add the text in the item's selected text in the Menu Editor outline.
- ❑ Edit the ShortCut property of the selected item by using the Property Editor.

To enter a shortcut for the Copy command:

1. In the Menu Editor window, double-click the Copy item.

   The Property Editor opens, displaying the properties of Copy.
2. Set the ShortCut property to the ^ (ins).
3. . . .

Even if the introduction to a list refers to the number of items in the list, do not use an ordered list unless the sequence or priority of the items is important. Just because you acknowledge that a list has two items, for example, doesn't mean that you must number them.

Keep list items parallel, to avoid the confusion in the following help topic:

Original

Help

If the new file has the same name as an existing file, do one of the following:

- ❑ Specify the REPLACE option, to replace the existing file.
- ❑ Give the file a new name (RENAME option).
- ❑ By indicating IGNORE, stop processing the file.

Revision

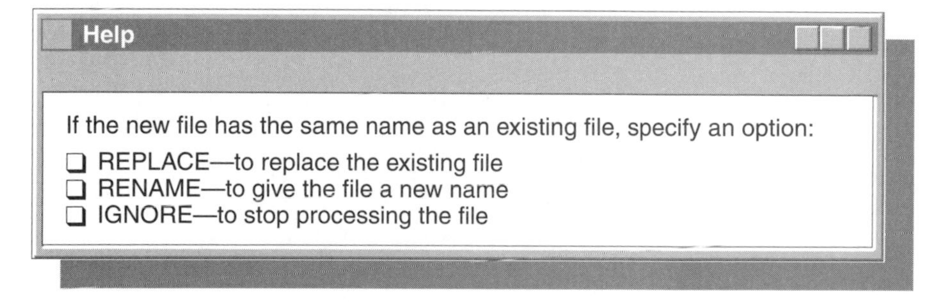

At first glance, the two help topics look alike: each is a list with three items. However, in the original topic, each item focuses on an action only superficially; users might puzzle over unexpected turns in the list items. In the revision, the items focus on the options and are consistent; users can predict what's coming. The items in the list have a similar structure, which is called *parallelism*.

As this example also shows, you need to consider how the introduction to the list relates grammatically to the list items. Make the introduction as specific as possible, moving back into the introduction the information that is common among the list items. For translation purposes, the best technique is to make the introduction a sentence rather than to make it grammatically dependent on the list items for an object or verb, for example. You can't be sure that other languages can break the flow at the same place and still make sense in both the introduction and the list items.

In terms of both the introduction and the punctuation of the list items, avoid treating a list as a vertical version of a horizontal list, as in the following list:

Original

The InfoCenter can be installed:

- On a server, for access from a browser on each user's computer, or
- On a user's computer, for access only from that computer.

This sentence needs more changes to take advantage of a bulleted list and clearly present the relationships between the introduction and each list item.

Revision

The InfoCenter can be installed in either or both of the following configurations:

- On a server, for access from a browser on each user's computer
- On a user's computer, for access only from that computer

The items in this list are alternatives, but they are not mutually exclusive. Installing the InfoCenter in one place doesn't mean that users can't also install it in the other place. They just don't need to install it in both places.

In the introduction to a list, clearly indicate how the list items relate to each other. They might have any of the following relationships:

- Alternatives where a specified number (but not all) apply; an implied *or* joins the items.
- Alternatives from which one or more (could be all) apply; an implied *or* joins the items.
- A group where all the items apply; an implied *and* joins the items.

The list items can then stand on their own.

Segment information into tables

Like lists, tables can effectively present similar technical information in a similar way.

Spread information across the columns. Some tables lump too much information in one column, often the last column, as in the following table:

Original

Option	Default	Comments
CalStackSize	10	CalStackSize=15 Sets the number of displayable items in the combination box for the Inspector Call Stack
CreationMode	ControlArray	CreationMode=AskUser—User confirmation CreationMode=ControlArray—No warning CreationMode=IndividualControl—Group created from Toolbox
UndoStackSize	10	UndoStackSize=# For example, a value of 5 limits **Edit —> Undo** and **Edit —> Redo** to five actions each

The last column of the original table is a catchall for different kinds of information.

Revision

Option	Default	Example	Explanation of example
CalStackSize	10	CalStackSize=15	Sets the number of displayable items in the combination box for the Inspector Call Stack
CreationMode	Control Array	CreationMode=AskUser CreationMode=ControlArray CreationMode=IndividualControl	Asks for user confirmation No warning needed Group created from Toolbox
UndoStackSize	10	UndoStackSize=5	Limits **Edit —> Undo** and **Edit —> Redo** to five actions each

The revision puts the examples in one column and the explanation of the examples in another column.

Avoid using tables when the information that you want to present is undifferentiated. Separate the common information and put it in text rather than have a column with the same entries. The following table shows that most of the information for the items in the first column is the same:

Original

Data type	Passed by reference	Passed by value
Integer	x	x
Long	x	x
Single	x	x
Double	x	x
Currency	x	x
String	x	x
Array	x	
Null pointers	x	
User defined	x	x
Variant	x	x
Any	x	
Class	x	

Revision

All data types can be passed by reference. All data types except the following data types can be passed by value: array, null pointers, any, and class.

Unless you have more information to add to the original table or you need to list the data types in one place, the information in the table might be better expressed in text, as in the revised passage.

Look for similarities among topics and parts of topics so that you can use elements such as lists and tables to present the information. Users can then immediately recognize the similarities in the information.

Use technical terms only if they are necessary and appropriate

Whenever you can reasonably avoid introducing a specialized term, do so. Unless a new term really helps users, it is just one more thing for them to learn. To use technical terms only if they are necessary and appropriate, follow these guidelines:

- ❑ Decide whether to use a term.
- ❑ Use terms consistently.

Decide whether to use a term

When deciding whether a term is necessary and appropriate, consider these factors:

- ❑ Does a suitable term for the thing that you want to name already exist?

  If a widely used and accepted term does exist, use it. You need not coin a new term, even if you're tempted to replace a technical term with a nontechnical term.

- ❑ Do you intend to use a term only once or twice? If so, you might be able to use a descriptive phrase instead.

  Do not create an acronym unless you will use it often, and then only if the phrase it stands for is unwieldy.

Are all the terms necessary in the following passage?

Original

> Each keyword describes one aspect of a program failure. A set of keywords, called a *keyword string*, describes a specific problem in detail. Because you use a keyword string to search a database, a keyword string is also called a *search argument*.

Revision

> Each keyword describes one aspect of a program failure. You use a set of keywords, called a *keyword string*, to describe a problem in detail, search a database for a similar problem, and obtain a fix for the problem.

The original passage introduces a term, *search argument*, that the user can probably get along without. The revised example uses generic words and focuses on the uses of the keywords rather than on adding to the user's vocabulary. Depending on the experience of the audience, you might also drop *keyword string* as unnecessary.

Use terms consistently

If you decide that a term is needed, carefully choose it, introduce it, and use it. If the general literature in your field uses two or more terms for the same thing, pick one and use it consistently. In some writing, you are encouraged to use different words for the same thing, for variety. In technical writing, however, more than one term for the same thing causes confusion. Two terms suggest two different things.

Consider the following sentence:

Original

> The Invoice class contains the LineItem class. The AddLineItem method of the Invoice class accepts only objects of the LineItem type.

The use of *type* instead of *class* in the original sentence could cause users to wonder whether the LineItem type is different from the LineItem class.

Revision

> The Invoice class contains the LineItem class. The AddLineItem method of the Invoice class accepts only objects of the LineItem class.

You might decide that mentioning an alternative term with which users are familiar will help them understand a new concept. In that case, be sure to clearly identify the one term that you intend to use thereafter. For example, you might deal with the similarity between *containment* and *aggregation* by writing:

> *Containment* (also known as *aggregation*) refers to the relationship in which a class is part of another class.

You can also add the alternative term to the glossary and index with a *See* reference to your chosen term.

Consider whether a term that you are considering suggests what it means. If it suggests nothing, users must learn it by rote. If it suggests something that is different from what it means, users will be confused.

Carefully choose terms that indicate distinctions among things. For example, a better contrasting term for *physical records* might have been *nonphysical records* rather than *logical records*, which suggests *illogical records* as its opposite. Physical records and logical records are common terms today and should not be changed. However, new users of these terms might understand them more easily if the distinction between them were more apparent in the words.

Be sensitive to whether a term is jargon. *Jargon* is a term that is used by people in a particular area of expertise. Such a term might pervade your project or your workplace, for example. A term that is jargon can take any of several forms, such as:

- ❑ Play off another word, as *automagically* plays off *automatically*.
- ❑ Have a different meaning from what's generally expected. For example, *hit* and *check* have a specialized meaning when used to refer to an error in software or hardware.
- ❑ Truncate a standard term (such as *nuke* from *nuclear*).
- ❑ Elongate a standard term (such as *ad-hockery* from *ad hoc*).

Avoid using jargon unless it is appropriate for your audience and no other word fits the situation.

If you are new to a project, you might be more sensitive to problems with terminology than are people who are familiar with the project. However, you should consult more experienced colleagues before changing the terminology in existing information.

Define each term that is new to the intended audience

In all writing, the transactions between the writer and the user of the information depend on the vocabulary that they have in common. When users first encounter a new subject, it might seem like a foreign language, with its own vocabulary and rules for combining the concepts that the terms express. For example, native speakers of English who have never developed a computer application might feel as lost as if they were in a foreign country the first time that they try to use an application development tool.

Some terms might be familiar, but their definitions might differ from what users expect. *Object* and *class*, for example, are familiar words, but they have special meanings in object-oriented programming.

Define terms that are new to users or are used differently from what users expect. Such terms include acronyms and other abbreviations. In the text where you first use a new term, explain the term in a meaningful way and highlight it. Try not to use the term before that.

To help users recognize the terms and definitions that you use, highlight new terms. In online information, you can make the definition available through a link. Pop-up help (or bubble help) is especially effective for presenting definitions because users can stay in the same topic while they read a definition. You can also have links from acronyms to the terms that they represent. However, users can probably understand an acronym more easily in online information if you use it only after you give the term in the same topic or if the acronym is commonly understood; you can't be sure that the user has seen another topic that explains the acronym.

In printed information, include a definition where you first use a new term and put the term in the glossary and in the index. If a term has an acronym or abbreviation that you use, spell out the first occurrence of the term, and follow it with the acronym or abbreviation in parentheses.

Including an example in the definition also helps users understand terms and makes the definition concrete. The following definition lacks an example:

Original

attribute	An intrinsic characteristic of a class or object. An attribute can have a value.

Revision

attribute	An intrinsic characteristic of a class or object. An attribute can have a value. For example, the attributes of a driver for an insurance application might include birth date, gender, and traffic violations. Values for these attributes might be 05/03/1978, female, and 0.

The original definition is abstract and hard to understand. The revised definition makes more sense because it draws on common experience to clarify the meaning of *attribute*.

In definitions, try to use terms that are familiar to the user. If you include unfamiliar terms, define them in the glossary or in context. When you use terms that you must define, be careful that the definitions aren't circular. Circular definitions, as in the following passage, are frustrating for users:

Original

Index entry	A ***key*** and a ***pointer*** paired together.
key	One half of the pair that makes up an ***index entry***.
pointer	One half of the pair that makes up an ***index entry***.

Revision

index entry	A ***key*** and a ***pointer*** paired together.
key	One or more consecutive characters taken from a data record; used to identify the record and establish its sequence.
pointer	An address or other indication of location.

The original definitions of *key* and *pointer* merely paraphrase the definition of *index entry*. The revised definitions are informative and are independent of each other.

Make sure that new terms are compatible with each other and with existing terms. If you define two terms separately, define the combination only if it has a special significance. For example, you don't need to define *class attribute* if you've defined *class* and *attribute*. However, you'd probably define *data type* rather than *data* and *type* because of their widespread use together.

In sum

Clarity in technical information covers a broad range of issues, most of which you can resolve at the level of words and sentences. You need to be aware of possible meanings that someone might read into what you write, other than what you intended. This awareness is particularly important when translators and nonnative speakers might use your information.

Use the guidelines in this chapter to ensure that technical information is clear to your audience. Refer to the examples in the chapter for practical applications of these guidelines.

When you review technical information for clarity, you can use the checklist on page 147 in two ways:

- ❑ As a reminder of what to look for, to ensure a thorough review
- ❑ As an evaluation tool, to determine the quality of the information

Judging by the number and severity of items that you found, decide how the information ranks on each guideline for this quality characteristic. You can then add your findings to "Quality checklist" on page 387, which covers all the quality characteristics.

Although the guidelines are intended to cover all areas for this quality characteristic, you might find additional items to add to the list for a guideline.

Guidelines for clarity	Items to look for	Quality rating
Focus on the meaning.	• Sentences are concise. • Paragraphs are focused. • Verbs are precise. • Meaningless words are not used. • Only necessary modifiers are used.	1 2 3 4 5
Avoid ambiguity.	• Words are easy to translate. • Antecedents of pronouns are obvious. • Modifiers are placed appropriately. • Strings of nouns are short. • Expressions are positive. • The syntax of sentences is clear.	1 2 3 4 5
Keep elements short.	• Expressions are concise. • Words are direct. • The length of lists is appropriate for the content and medium.	1 2 3 4 5
Write cohesively.	• Transitions are smooth. • Lists, tables, and examples are integrated.	1 2 3 4 5
Present similar information in a similar way.	• Types of lists are appropriate for the information. • List items are parallel in structure. • Tables are well segmented.	1 2 3 4 5
Use technical terms only if they are necessary and appropriate.	• Terms that are used are necessary. • Terms are appropriate. • Use of terms is consistent. • Jargon is not used.	1 2 3 4 5
Define each term that is new to the intended audience.	• Terms are defined. • Acronyms and abbreviations are defined. • Definitions are easy to understand.	1 2 3 4 5

Note: The scale for the quality rating goes from very satisfied (1) to very dissatisfied (5).

Chapter 6

Concreteness

You look at me as if you're in a daze
It's like the feeling at the end of a page
When you realize you don't know what you just read. —Missing Persons

Concreteness in technical information communicates what people can see, hear, smell, taste, or touch—not just what they can think or imagine. It deals with material things and physical actions in the real world, the nitty-gritty.

Many technical subjects, such as relational database access, inheritance among classes, and communication protocols, are abstract. Users cannot experience them directly, as they can see and touch the concrete blocks in a building, for example. Few users enjoy wandering among concepts and grasping at new terms without being able to connect them to something real in their experience.

Because people can more easily understand the concrete, a good writer relates the abstract to the concrete. You must be careful to give only as much conceptual information as really helps users understand their tasks and to present such information concretely. For example, information about why a particular implementation was chosen might be too theoretical for users.

Language and presentations can run the gamut from the abstract (or theoretical) to the concrete (which can itself be general or specific), as shown in Figure 7 on page 150.

Figure 7. Examples of the continuum from the abstract to the concrete.

Abstract ←			→ Concrete
financial institution	bank	Federated Bank	Federated Bank, San Jose
computer system	software	groupware	Lotus Notes groupware
hierarchy	pyramid structure	organization chart	organization chart, ABC Company
"A function of the compiler is to check syntax."		"The XYZ compiler checks syntax."	example of syntax messages for a code sample
"A database is a set of tables, in which each table is a set of relations."		"A database is like a file drawer, in which each folder holds related papers."	

In addition to meaning "theoretical, not practical," abstract is sometimes equated to "abstruse, or not easily understood." Some techniques that you use to make information clear (in the clarity chapter on page 103) can also help lessen abstractness. This chapter about concreteness focuses on more techniques, as shown in Figure 8 on page 151.

Any of several factors probably keep writers from writing concretely:

- You need experience with a subject to know what examples or scenarios to use, creativity to recognize the possibilities for making information more concrete, and effort to make the changes or additions. Often writers do not have time to become comfortable with a subject, or the subject is complicated.
- The items that make information more concrete are often regarded as extra work, something to do after you have written the basic information. Writers might choose not to include examples and specifics in a plan or outline. Examples that are missing might not be so obvious during a review as a missing step or task. Ironically, users do miss examples because users depend on being able to see something like what they want to do.
- Writing concretely can be risky in that something concrete is more likely to be inaccurate or dated than something abstract. You need to be willing to take this risk and work to reduce it.

You can use the means of achieving concreteness to define, illustrate, compare and contrast, and even enliven information. Concrete elements such as examples, scenarios, and similes are all means of avoiding the dullness that many people consider typical of technical information.

Figure 8. Concreteness offers many ways to counteract abstractness.

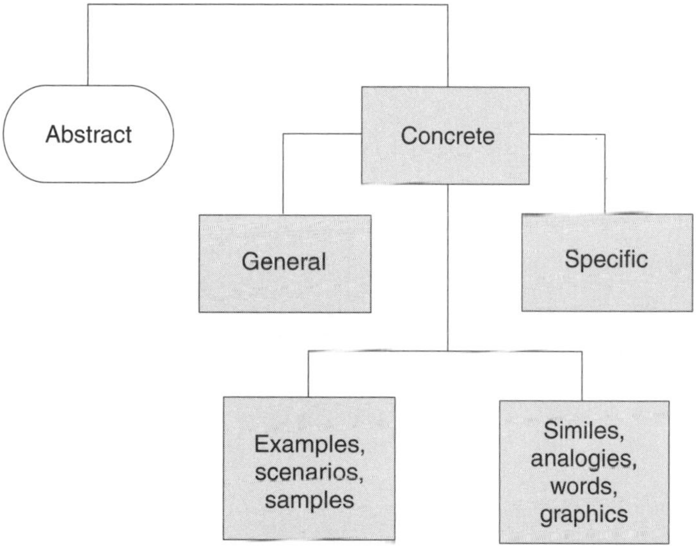

As with Figure 8, illustrations even of abstract subjects help give users something to see so that they can visualize relationships.

Because of the importance of examples and scenarios to users, this chapter emphasizes these means of achieving concreteness. This chapter also focuses on the quality characteristics (particularly task orientation, accuracy, clarity, and retrievability) that make examples and scenarios effective.

To make information concrete, follow these guidelines:

- ❑ **Choose examples that are appropriate for the audience and subject.**
- ❑ **Use focused, realistic, accurate, up-to-date examples.**
- ❑ **Make examples easy to find.**
- ❑ **Make code examples easy to adapt.**
- ❑ **Use scenarios to illustrate tasks and to provide overviews.**
- ❑ **Set the context for examples and scenarios.**
- ❑ **Relate unfamiliar information to familiar information.**
- ❑ **Use general language appropriately.**

Choose examples that are appropriate for the audience and subject

An *example* is a representative of a set of things. It can be typical of the set or atypical. Atypical examples and examples from outside the set (sometimes called *nonexamples*) help to show the boundaries of what is included in the set and what is not included, as shown in Figure 9.

Figure 9. Different kinds of examples help define a domain.

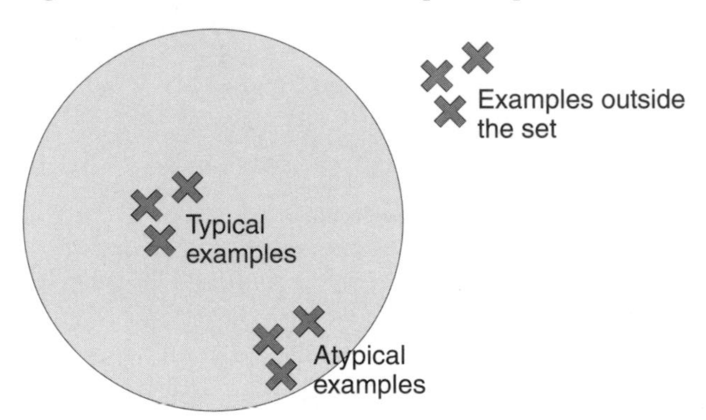

Examples are probably the purest form of concreteness—and the most useful form of information for users. No matter how many you supply, users will probably ask for more. Unfortunately, providing good examples takes time, and you need to decide what examples you can best provide.

The following guidelines help you choose examples that are appropriate for the audience and subject:

- ❑ Consider the level and needs of users.
- ❑ Use examples appropriately in conceptual, task, and reference information.

Consider the level and needs of users

Users should be able to understand an example and apply it to their circumstances. You might need to have examples for different skill levels and from various subjects. The users most likely to need examples are the ones who:

- ❑ Have little experience with the subject or with related subjects
- ❑ Tend to use a product by strictly following instructions

As with many guidelines in this book, the key principle is to *know your user*. To understand the skill levels and get ideas for examples, check with your users, and check journals and documentation for similar subjects or products or for similar uses of different products. By observing users during usability testing, you can also get ideas for examples or for places in the information that most need examples.

One question that sometimes comes up when writers plan for examples is whether to unify the examples around a particular situation. For example, you might base your examples on applications that are used by a fictional hotel or marketing firm. The main advantage of such examples is that testing can be easier if the examples come from a large application. The main disadvantage is that unified examples can require users to be familiar with introductory information or with previous examples in the set. This approach can be fine for a tutorial where users are likely to do the lessons in sequence, but is less appropriate for information that users are unlikely to read from start to finish. Also, unless your audience is homogeneous, a variety of situations for examples can give you a better chance of appealing to more users.

Consider also the need to appeal to an international audience. Avoid situations and terms that are specific to a particular culture, as in the following table of examples:

Original

Examples of valid names	Examples of not valid names	Rule for valid name
inch	3inch	Name must start with letter or underscore.
retirement_account MayJune	retirement account May/June	Name must not use special characters (other than underscore) or blanks.

Revision

Examples of valid names	Examples of not valid names	Rule for valid name
centimeter	3centimeter	Name must start with letter or underscore.
savings_account MayJune	savings account May/June	Name must not use special characters (other than underscore) or blanks.

The original table of examples uses terms such as *inch* and *retirement account* that might not be common outside the United States. The revision replaces these terms with ones that are more likely to be international.

This table of examples also shows the use of examples that are outside the acceptable set. Such examples help define what is acceptable.

Use examples appropriately in conceptual, task, and reference information

You can use examples in different types of technical information to make them easier to understand. In conceptual information, you can use examples to help define concepts. When you define a noun, you need to specify the class to which the thing belongs and then the qualifiers that distinguish it within that class, as in the following definition:

Original

> In the library system, a data format is a predefined method or set of instructions that is associated with an object. Data formats define the type of information that an object contains and how to work with that object.

This definition classifies a *data format* as "a predefined method or set of instructions" provided that the method or set "is associated with an object." The second sentence adds more information, but the definition still seems abstract and unrelated to something that users do. Consider the following revision of this definition:

Revision

> In the library system, a data format is a predefined method or set of instructions that is associated with an object. Data formats define the type of information that an object contains and how to work with that object.
>
> For example, if you create and save an object as a Word file, you can import the file, associate it with the data format for Word, and store it as an object in the library system. You can then work with the file in its own format.

The revision adds an example that relates the concept of a data format to the task of creating and saving an object as a Word file. By seeing an application of the concept, users can more easily understand the meaning of this concept.

You can also use graphics to make the meaning of a concept more concrete, as in the following passage:

Original

> Standard Generalized Markup Language (SGML) is a rich language that describes markup languages, particularly those for exchanging, managing, and publishing electronic documents. Rather than specifying particular tags, it defines rules and guidelines for identifying tags.
>
> Two main uses of SGML have evolved: applications and subsets.
>
> - HyperText Markup Language (HTML) is an application of SGML that is specifically for displaying documents on the Web. HTML tags are established by a standard and are hard to change apart from the standard.
> - Extensible Markup Language (XML) is a subset of SGML that facilitates making tags to fit the content of one's information.
>
> Extensible HyperText Markup Language (XHTML) extends HTML as an application of XML.

Revision

> Standard Generalized Markup Language (SGML) is a rich language that describes markup languages, particularly those for exchanging, managing, and publishing electronic documents. Rather than specifying particular tags, it defines rules and guidelines for identifying tags.
>
> Two main uses of SGML have evolved: applications and subsets.
>
> - HyperText Markup Language (HTML) is an application of SGML that is specifically for displaying documents on the Web. HTML tags are established by a standard and are hard to change apart from the standard.
> - Extensible Markup Language (XML) is a subset of SGML that facilitates making tags to fit the content of one's information.
>
> Extensible HyperText Markup Language (XHTML) extends HTML as an application of XML.

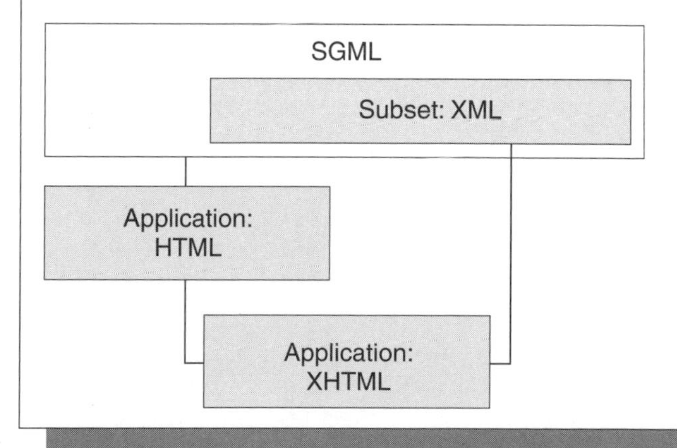

The original passage describes SGML, HTML, XML, and XHTML, relating them to each other. However, the illustration in the revision makes the relationships much easier to understand.

Examples are also important in task information, both within a step and for the overall task. Within a step, users appreciate examples of what they must enter and examples of what they can expect the system to return. Consider the following set of steps:

Original

1. Identify the version and release of the ProgLang product that you use. To find the version and release, compile a program by using the LIST option. The version and release are shown as the last item at the top of the listing output.
2. Search FAQs or hints and tips, and select the operating system and release of the product that you use.
3. For the problem that you encountered, enter keywords and the identifier for any error messages that you received while building or running the application.

Revision

1. Identify the version and release of the ProgLang product that you use. To find the version and release, compile a program by using the LIST option. The version and release are shown as the last item at the top of the listing output. For example, you might see the following line:

   ```
   CC 5648-A25 PROGLANG 2.1.2
   ```
2. Search FAQs or hints and tips, and select the operating system and release of the product that you use.
3. For the problem that you encountered, enter keywords and the identifier for any error messages that you received while building or running the application. For example, you might enter:

   ```
   ABENDOC4 XYAZDS0084-S
   ```

The original version makes the steps seem more difficult than they are because it gives no examples. In the revised version, the example of output in the first step and the example of input in the last step help users know what to expect.

You can't provide an example for every step, and you shouldn't need to. If the user can see the actual interface while doing a step, such as the second step in the preceding example, the user probably doesn't need a screen capture of the interface. However, in some situations, inserting a screen capture can more clearly show what to do. For information about when and how to include screen captures, see the visual effectiveness guideline "Avoid illustrating what is already visible" on page 280.

Because users often use a trial-and-error approach and turn to the information only when something goes wrong, users also appreciate examples of common errors along with task information.

Given time constraints, providing an example for every task is probably impossible and perhaps not desirable from the users' point of view. How do you decide what tasks most need examples? You can consider certain characteristics of the tasks, as shown in Figure 10.

Figure 10. Criteria to consider when you select examples and scenarios.

Importance and newness ↑ / Difficulty and frequency →		
	Least difficult Least frequent New Most important	Most difficult Most frequent New Most important
	Least difficult Least frequent Familiar Least important	Most difficult Most frequent Familiar Least important

Difficulty and frequency

Focus particularly on the following kinds of tasks (as shown in the upper right quadrant of the figure):

- ❑ Critical to using the product (most important)
- ❑ New or changed
- ❑ Frequently done but difficult

Consider whether the task example used in the following passage is appropriate:

Original

> You can do many tasks from the InfoShopper main window. You can click Color to change the color of any of the windows. For example, click the red push button to make the window red.

The original passage tries to show how easy the product is to use by presenting the ease of changing the color of the window. However, this example isn't pertinent to the user's main tasks.

Revision

> You can do many tasks from the InfoShopper main window. You can list the available merchandise in a category by clicking the picture that corresponds to your category. For example, to see a list of jewelry, click the picture of a ring.

The revised passage uses the task of listing merchandise to show how easy the product is to use. This task is something that users can relate to. In addition to learning how easy the product is to use, the users learn how to do a relevant task. Examples help users learn to apply information.

Examples of programming statements such as methods and commands give users valuable information about syntax and usage. Use such examples liberally in both task information and reference information.

The following passage requires users to read the explanation of the syntax carefully and provides no example for users to follow:

Original

> In MathML, use the <apply> element to indicate the operator to apply to the arguments in a mathematical expression. The <apply> element takes the place of parentheses.

Revision

> In MathML, use the <apply> element to indicate the operator to apply to the arguments in a mathematical expression. The <apply> element takes the place of parentheses, as in the following example:

```
<apply><times/> <ci> a </ci> <ci> b </ci> </apply>
```

> This tagging resolves to (*a* x *b*).

The revised passage shows users what they need to type if they specify the tags. The passage provides both an example of the syntax and the equivalent result, in case the users are unfamiliar with the tags in the example.

You might also provide a sample (a large set of code or data for using a product) such as:

- A sample database so that users can practice using your product to populate a database

- A sample application so that users can practice using your product to test an interface

Samples can take the place of descriptive information and give users a chance to learn by doing, as in the following pair of examples:

Original

> Before you use the InfoDBase product, read the manual. It is important that you understand all the tasks before you use InfoDBase with your database.
>
> . . .
>
> Sample SQL is provided in Appendix A. You can copy the samples from the appendix if you want to use them. For more information about how to write SQL statements, see the *SQL Operator's Guide*.

Revision

> You can practice performing the tasks in this chapter before you try them on your own database. Use the sample database, which is provided with the InfoDBase product, instead of your own database until you are comfortable with the tasks. To access the sample database, do these steps:
>
> . . .
>
> You can use the SQL that is provided in the SAMPLES directory for steps 3, 4, and 5, or you can write your own SQL.

The original passage provides little help to users. It might actually frustrate them by setting such a high first hurdle as reading the whole manual, or alarm them by implying serious consequences from trying to learn by trial and error. The revised passage indicates a friendly, task-oriented product that provides users with helpful samples that they can use to do their tasks.

Use focused, realistic, accurate, up-to-date examples

To make examples serve their purpose forcefully, writers sometimes simplify examples. Removing extraneous parts from an example helps users focus on the point of the example. The first part of this guideline is similar to the clarity guideline "Focus on the meaning" on page 105. In the original passage that was improved for that guideline, the example contained more information than was needed to make the point about the usefulness of digital images in a security system.

Writers seldom have real examples, especially code examples, and permission to use them. Realistic (rather than real) examples are one possibility for filling the need for examples. If you are dealing with a complex system, choose examples that show increasing levels of complexity rather than trying to show too much all at once.

Regardless of whether an example is real or realistic, the parts of an example that users depend on must be accurate. Inaccurate code examples can be worse than no code examples. Be sure to test code examples, even though this testing requires extra effort. Code fragments and stand-alone examples probably need scaffolding (code that simulates the function of nonexistent components) to enable testers to test them. Sometimes reading such examples is the most that testers can fit into the testing schedule. Taking lines of code from a larger application is much easier, because you can assume that the parts are correct if the whole application works.

Try to find the error in the following passage:

Original

This statement gets the value of the attribute price, figures 5% of its current value, and stores the result in the variable $commission:

```
getMult('price', 0.50, $commission)
```

Revision

This statement gets the value of the attribute price, figures 5% of its current value, and stores the result in the variable $commission:

```
getMult('price', 0.05, $commission)
```

The original example missed 5% by a factor of 10. A user would probably decide that the text was more likely to be right than the actual code. A tester reading the code example in the original passage would probably also find this error, which the revised example corrects. However, such errors are hard to find if you are reading for format rather than for sense.

A code example can become outdated and inaccurate if it contains functions that are no longer supported. Code examples and other kinds of examples can become outdated if they are items from popular culture or numbers (such as prices and years). If you must include years, try to make them far enough in the future that they won't soon seem outdated. Consider how long the information must last.

To reduce the work that is needed to maintain examples, remove items that are likely to show their age quickly. Consider what item is likely to become dated in the following example and how you might change it:

Original

> This statement creates an object, an instance of the class car. Unless a later statement provides a different name, the product will use the name myBronco for the object and assign the name to the variable *mycar*:
>
> ```
> car.new('myBronco', mycar)
> ```

Revision

> This statement creates an object, an instance of the class car. Unless a later statement provides a different name, the product will use the name myFord for the object and assign the name to the variable *mycar*:
>
> ```
> car.new('myFord', mycar)
> ```

The original example uses a model name for a car, which is more likely to change than the manufacturer name, which the revised example uses.

Whatever examples you use, be sure to make the explanatory text match the example.

Make examples easy to find

Examples can't help users much if they're hard to find. Guidelines from organization, retrievability, and visual effectiveness apply to examples, whether the examples are long or short.

To make examples easy to find, follow these guidelines:

- ❑ Use visual cues to indicate where examples are.
- ❑ Make examples part of the user interface.
- ❑ Make clear where examples start and stop.

Use visual cues to indicate where examples are

Rather than burying text examples in a paragraph, set them off, perhaps by starting a new paragraph or a new line, adding a heading or label, using a list, or adding a column to a table.

Users who scan are likely to miss the examples in the following paragraph:

Original

InfoProduct maintains information about attributes that are unique to the image data type, as well as information about attributes that are common across data types (for example, common across the image and audio data types). For example, the product maintains information about the width, height, and format of an image, as well as information about common attributes such as the identification of the person who imported the object into the database or who last updated the object.

Revision

InfoProduct maintains information about two kinds of attributes:

- ❑ Attributes that are unique to the image data type, such as the width, height, and format of an image
- ❑ Attributes that are common across data types, such as the identification of the person who last updated the object (whether an image, audio clip, or some other type of object)

The first passage repeats too much information for the sake of a few examples. The repeated information overwhelms the examples and makes them hard to find and understand. The revision uses a list to avoid repeating information and to make the examples stand out.

Using a column in a table can draw more attention to examples. The tables on page 139 and page 153 show the use of columns to set off examples.

Make examples part of the user interface

Usability research shows that users prefer information that is closest to the product interface—even a click and scroll or function key away can be too much. The information needs to be part of the words and choices in the product interface. You can assimilate some information into the interface as examples, as in these cases:

- Where you might use contextual help for an entry field, use instead an example that shows both the default and the format.
- Use combo boxes (a text box with an attached list box) to show users their choices for constructing a programming statement.
- Provide an example in a message.

Look for places where you can replace text with examples rather than just add examples. Figure 11 shows the use of help to inform the user of the format that is expected for the value.

Figure 11. A message provides information about the expected format for the value.

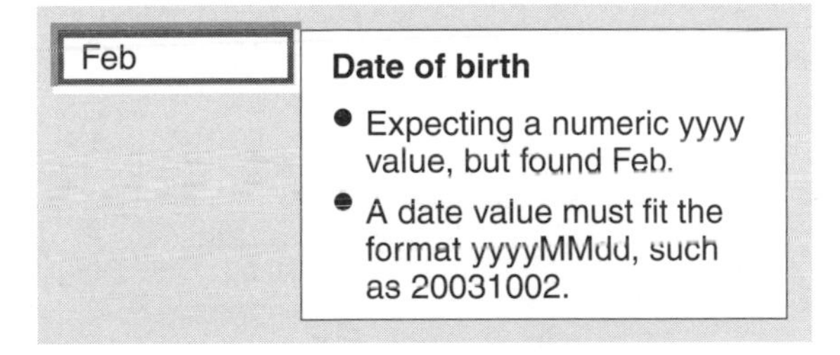

Make clear where examples start and stop

When you add a heading or label to set off one or more examples, the information under the heading or label should relate only to the example or examples. Do not include general or substantive information. A user who chooses not to read an example should not also miss the only place where certain information is given.

Consider the arrangement of information in the following concept topic:

Original

Storage classes

A storage class is a logical entity through which you can define the retention period for a set of objects in the library system.

If you increase the retention period (for example, from 10 to 20 days), all objects that are associated with the storage class and any objects that have been added to the storage class remain for the new retention period of 20 days.

Example: If you decrease the retention period from 20 to 10 days, the library system does not migrate the objects that are associated with the storage class until they reach the previous retention period of 20 days. After the new retention period of 10 days, the library system migrates any new objects that are associated with this storage class. Objects remain on the last storage class indefinitely, regardless of the retention period that you specify, to prevent the system from migrating objects off the last storage class and deleting them.

The system moves the objects from one storage class to another based on the order and retention period that you assign to each storage class.

Revision

Storage classes

A storage class is a logical entity through which you can define the retention period for a set of objects in the library system.

The library system moves objects from one storage class to another based on the order and retention period that you assign to each storage class. Objects remain on the last storage class indefinitely, regardless of the retention period that you specify, to prevent the system from migrating objects off the last storage class and deleting them.

Example: If you increase the retention period from 10 to 20 days, all objects that are associated with the storage class and any objects that have been added to the storage class remain for the new retention period of 20 days. If you then decrease the retention period from 20 to 10 days, the library system does not migrate the objects that are associated with the storage class until they reach the previous retention period of 20 days. After the new retention period of 10 days, the library system migrates any new objects that are associated with this storage class.

The original topic mixes general information with examples both within the first sentence and within the paragraph labeled "Example." This arrangement is very confusing, because the headings do not match their contents. Users are likely to wonder which parts of the information are intended to be the example and which parts apply beyond the example. Users must read the text carefully to sort out the general information from the example.

The revision clearly separates the general information from the example, making the example much more effective. The example paragraph now meets the users' expectations for the information there.

Using headings and labels to set off examples has these advantages:

- Making examples easy to find in the help or on a page
- Showing the examples in the navigation pane (as in Figure 12) or in a table of contents

Figure 12. The examples category in the navigation pane makes the examples easier to find.

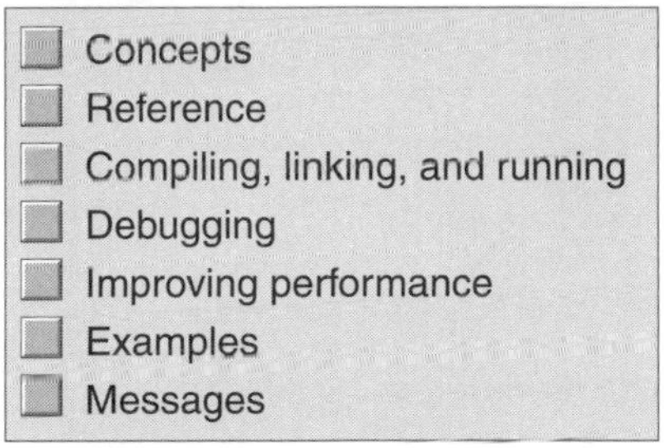

- Concepts
- Reference
- Compiling, linking, and running
- Debugging
- Improving performance
- Examples
- Messages

If you are chunking information into types such as tasks, concepts, and reference, you can also chunk long examples and link to them from one or more tasks.

Make code examples easy to adapt

Code examples help users do their own work because users can copy the example and adapt it to their needs.

Examples in software information often contain code and commands. A C^{++} routine does not describe C^{++} code; it *is* C^{++} code for a particular purpose. When describing the C^{++} language, a good writer uses examples of C^{++} code.

Some code examples are hard to understand and use because they use ambiguous variables or lack explanatory comments, as shown in the following example:

Original

```
/* */
parse arg s n'('p
if s = '' then s = 's'
if n = '' then n = '<n>'
if p = '' then p = '<p>'
'extract /line'
'bot'
'i 's'--'strip(n)', 'p
':'line.1
'no' substr(date(month),1,3)
```

Revision

```
/* Add a salutation to a note and change the notebook */
parse arg salutation name '('phoneNumber
if salutation = '' then salutation = 'thanks'
if name = '' then name = '<your name>'
if phoneNumber = '' then phoneNumber = '<and number>'
'extract /line'
'bottom'
'input 'salutation'--'strip(name)', 'phoneNumber
':'line.1
/* Call the no macro to change the notebook to the current month */
'no' substr(date(month),1,3)
```

The original code example is hard to understand because it has no comments and uses variables that are too short to suggest what they represent. The revision remedies these problems.

When you add comments to code examples, consider where they will be most helpful—probably not far from what they're explaining. Consider the following examples:

Original

The following example creates a user-defined function (UDF) named `map_scale` that calculates the scale of a map. Notice that the UDF identifies `map` as the data type to which it can be applied. The code that implements the function is written in C and is identified in the EXTERNAL NAME clause.

```
/* */
CREATE FUNCTION map_scale (map)
  RETURNS SMALLINT 's'
  EXTERNAL NAME 'scale!map'
  LANGUAGE C
  PARAMETER STYLE DB2SQL
  NO SQL
  NOT VARIANT
  NO EXTERNAL ACTION
```

Revision

The following example creates a user-defined function (UDF) named `map_scale` that calculates the scale of a map.

```
/*map is the data type to which map_scale applies*/
CREATE FUNCTION map_scale (map)
  RETURNS SMALLINT 's'
/*scale!map is the name of the C function that implements the UDF*/
  EXTERNAL NAME 'scale!map'
  LANGUAGE C
  PARAMETER STYLE DB2SQL
  NO SQL
  NOT VARIANT
  NO EXTERNAL ACTION
```

The original passage uses sentences outside the code to give information that is more clearly presented in the revised passage as comments in the code. Users can easily miss the information when it is separate from the code. Also, users can see the comments if they access the code online.

One good way to draw attention to particular parts of a code example is to put numbers on certain lines and corresponding numbered notes at the end of the example, as shown in Figure 13 on page 168:

Figure 13. A code example with annotations for items that users need to understand.

```
<task> 1
<title>Configuring InfoPlayer</title> 2
<taskbody> 3
<steps>
<step> 4
<cmd>Start InfoPlayer.</cmd>
</step>
<step>
<cmd>Click Options —> Preferences.</cmd>
</step>
<step>
<cmd>On the Transport page, click Automatically select best trans-
port.</cmd>
</step>
<step>
<cmd>Click Auto Configure, and then click OK.</cmd>
</step>
</steps>
</taskbody>
</task>
```

1. The element <task> is the highest-level container for this topic.
2. Every topic has a title.
3. The element <taskbody> contains steps, but it can also contain prerequisites for the task and postrequisites.
4. Each step must contain the element <cmd>, but it can also contain a step result.

When you add notes to a code example, you can highlight areas that users need to change or that might confuse users.

If your examples are large sections of code, provide them online with the product, rather than only in printed information. Users need to transcribe large code examples that are not online before they can copy or adapt the code.

Use scenarios to illustrate tasks and to provide overviews

A *scenario* depicts a series of events over time, usually around a fictitious but realistic set of circumstances. It shows one path through one or more products, a path that can be broad or narrow, high or low, vertical or horizontal. It can range from a high-level overview to a sequence of interface manipulations (sometimes accompanied by captured windows).

You use scenarios primarily either to teach (as in a tutorial) or to describe and motivate (as in marketing information). Unlike the lists of steps in a procedure, scenarios usually have a story that helps guide the actions. Consider the following introduction to a tutorial lesson:

Original

In this lesson you learn how to write an application to process data for more than one application server.

You can manage the transfer of data in one of the following ways:

- ❑ Use virtual memory as a buffer for holding a few rows of data.
- ❑ Use disk space as a buffer for holding many rows of data.
- ❑ Use multiprocessing to run each update in a different thread.

Revision

In this lesson you learn how to write an application to process data for more than one application server.

Suppose that the DoItRight Corporation periodically sends a copy of the employee table that is stored at headquarters to the regional branches. The application that you are writing needs to refresh the local copies. However, because a database manager can be connected to only one application server at a time, the application must manage the transfer of the data.

You can manage the transfer of data in one of the following ways:

- ❑ Use virtual memory as a buffer for holding a few rows of data.
- ❑ Use disk space as a buffer for holding many rows of data.
- ❑ Use multiprocessing to run each update in a different thread.

The original passage gives only the tasks, without a context in which to understand them. The revised passage puts the user in a realistic situation with a need to use the task information.

Consider the following passage from information for evaluating a product:

Original

> With a distributed relational database, you can benefit from accessing data as if it were located wherever you are.

Revision

> With a distributed relational database, you can benefit from accessing data as if it were located wherever you are.
>
> Consider a company with headquarters in one city and sales offices in many other cities. Each sales office has a relational database for its own transactions, and each month the office sends data to headquarters for the sales summary report. Using a distributed relational database that is a collection of the sales office databases, headquarters could access the sales data directly.

The original passage leaves to the user's imagination what the benefits might be. The revised passage gives a scenario of a particular benefit that users can apply to similar situations.

Some writers make the mistake of thinking they've made a scenario just by adding a thin story layer to what is still hard-to-understand information, as in the following topic:

Original

> **Connection ID usage**
>
> In this scenario, a two-party call exists between party A and party B. The Make_Call from party A to party B results in connection IDs being generated to represent their participation in the call. Party A then issues an Extend_Call program call to extend the call to party C. The Extend_Call places party B on hold and results in two new connection IDs being created: one for party A's connection in the new call and one for party C. To join all three parties in a conference call, party A issues a call to InfoConference specifying requesting party ID 1 and requesting party ID 2.

Revision

> **Conference call and connection IDs**
>
> Alice needs to set up a conference call with Subir and Calvin so that all three of them can have a discussion. Alice calls Subir, puts him on hold while she calls Calvin, and then connects Calvin into the call.
>
> The implementation of this conference call in terms of connection IDs looks like this:

Event	Connection IDs
Alice calls Subir	a1, b1
Alice extends the call to Calvin	a2, c1
Alice calls InfoConference	a1, b1, c1

The original scenario might represent a real situation inside the telephone software, but that's not where most people live or interact. Users might need to read this narrative several times just to understand whether the parties in it are people and what they are doing.

The revised scenario clarifies the situation and the relationship between what the people do and what the software does in response. Scenarios can emphasize actions and actors, without the imperatives that are usual in technical information. Writing scenarios is probably as close to narrative writing as writers of technical information get.

You need to create a scene with characters who have similar tasks or similar problems to those of your users. When you use scenarios, focus on tasks that users are likely to want to do. Scenarios that include interesting and relevant tasks are more helpful than those that focus on rarely used features. You might be able to get ideas for scenarios from the scenarios that usability engineers use for usability testing.

The scenarios that are used in this book as examples are necessarily brief. However, scenarios in technical information can run several pages long, depending on the number and complexity of tasks covered. Scenarios that deal with the use of more than one product are particularly likely to be long. For example, an entire book might provide scenarios for working with a group of products.

Set the context for examples and scenarios

Writers of technical information need to make explicit what's important about an example or scenario rather than just dropping it in and assuming that users will see in it something in particular. It might be obvious to you why you include an example or scenario or what you expect a user to get from it. However, users might not see what you want them to see in an example or scenario, as in the following passage:

Original

> To enter a ringtone in your cell phone by using a keypress sequence, do the following steps:
>
> 1. Select **Settings —> Ringtone Composer**.
> 2. Select **New Tone**.
> 3. Type the keypress sequence for the melody that you want. For example:
>
>   4 4 4 (1#) 2# 2# 2# (1)
>
>   Ignore the display on the screen. It reflects the composer's transposition of what you have typed.
> 4. Press **OK**. Name your melody and save it. The new melody will be listed with your other ringtones.

Someone who has not typed a keypress sequence before is not likely to understand this example. Even someone who is familiar with the symbols in a keypress sequence might like to know what song this sequence represents. Just writing "For example" is not enough of an introduction.

Revision

> To enter a ringtone in your cell phone by using a keypress sequence, do the following steps:
>
> 1. Select **Settings —> Ringtone Composer**.
> 2. Select **New Tone**.
> 3. Type the keypress sequence for the melody that you want. For example, the following sequence represents the opening bars of Beethoven's Fifth Symphony. Each number in a keypress sequence represents a note; the parentheses indicate longer notes, so you hold those keys down longer:
>
>   4 4 4 (1#) 2# 2# 2# (1)
>
>   Ignore the display on the screen. It reflects the composer's transposition of what you have typed.
> 4. Press **OK**. Name your melody and save it. The new melody will be listed with your other ringtones.

The revision includes an explanation of major features of the example that are important for users to know so that they can enter their own keypress sequence. The name of the song might also help some users recognize the notes in the sequence.

The introduction to an example or scenario should give users enough information that they can understand and apply the example or scenario to their own situation. They should be able to derive from the example the benefit that you intended.

Relate unfamiliar information to familiar information

Comparing the unfamiliar with the familiar makes users feel more comfortable with new information. Again, you need to understand your users to know what information is likely to be familiar and what information is likely to be unfamiliar to them.

Similes and analogies are the explicit description of one thing in terms of another. Words such as *like*, *as*, and *as if* usually introduce similes and analogies. These figures of speech add color and interest as well. Some users, for example, might have difficulty understanding what a database is. But when they read that tables in a database are like folders in a file drawer, they can understand the concept because they have used folders and file drawers.

However, similes and analogies can introduce unwanted associations and emotions. Therefore, when you use similes and analogies, make the basis of the comparison explicit. Avoid using metaphors, which are implied descriptions of one thing in terms of another.

Consider the following use of a simile to help explain a concept:

Original

A *class hierarchy* is a ranking of classes that shows their inheritance relationships. It looks like a tree.

Revision

A *class hierarchy* is a ranking of classes that shows their inheritance relationships. It resembles a tree because the classes that are lower in the structure are connected to classes that are higher up.

The original passage makes a vague assertion about the similarity of a class hierarchy to a tree. Various pictures of trees are likely to come to mind, but you don't know why they might look like a class hierarchy. The revised passage gives a reason for the suggested likeness. This situation could also be a good place to use a graphic to support the simile.

Analogies are similar to similes but are usually longer. Rather than relying on one area of similarity, they tend to involve a larger relationship. Consider how an analogy might improve the following passage:

Original

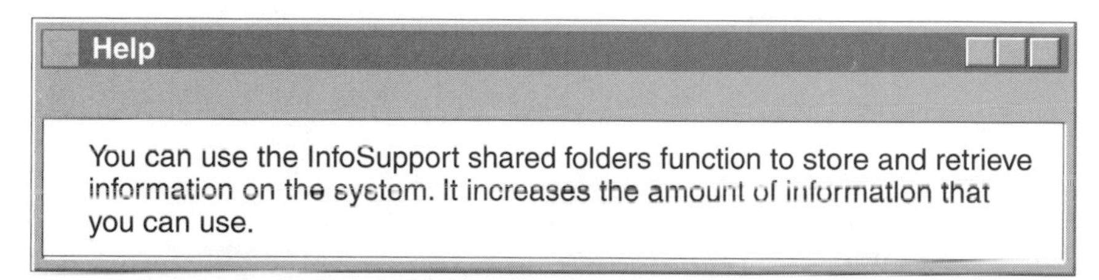

Revision

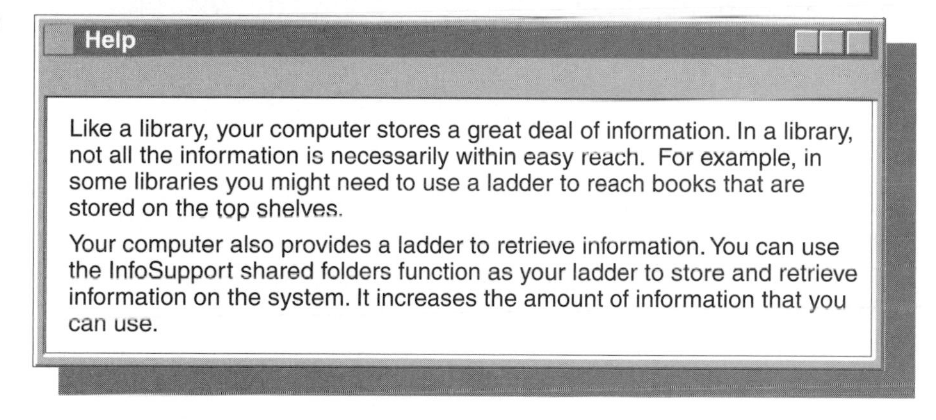

The original passage states the facts but lacks immediacy. The revised passage uses an analogy that many people can visualize and understand.

Often technical information deals with subjects that are new to at least some users. If a subject is well understood by most users, no gap exists in their understanding, but you could choose to use similes or analogies to enliven the information.

Use general language appropriately

Technical discussions include both general language and specific language. Users benefit from overviews, summaries, and topic sentences, which typically include general language. However, you need to weed abstractions out of general statements, and specific language should probably predominate.

The usual form for paragraphs in technical writing is to proceed from the general to the specific (deduction) rather than from the specific to the general (induction). Writers of technical information tend to present a principle or general description and then particular cases or exceptions rather than deriving the general statement from the particulars.

Consider the development of the following paragraph:

Original

> InfoLibrary does not enforce an explicit size limit for objects that are accessed and can support large objects to allow transparent object movement among servers. However, when a client application stores a file on an object server or retrieves a file from an object server, there is a relationship between the size of the object that can be stored or retrieved and the resources (disk and memory) available on the client workstation. When an object is stored or retrieved by the client from the object server, it is constructed in memory more than once as it is being processed on its path to and from file storage. If an object is too big for a given configuration of available memory, the required system resource is not available to create these copies. When this happens, a timeout occurs on the client.

Applying some clarity guidelines such as "Focus on the meaning" on page 105, "Avoid vague referents" on page 113, "Avoid long strings of nouns" on page 116, and "Write positively" on page 118 can help improve this passage. However, you can also think about improving it in terms of abstract, general, and specific language. This passage contains general statements that contain many abstractions such as "file storage," "size limit," "transparent object movement," "relationship," "resources," "given configuration," and "required system resources." These abstractions also help give the impression that the sentences in this paragraph are all at the same level of generality.

Nouns are more abstract than verbs, and pronouns are more abstract than the nouns that they refer to. The following revision eliminates the abstractions by using verbs and nouns appropriately:

Revision

> InfoLibrary supports moving objects of any size among servers. However, when a client application stores a file on an object server or retrieves it from there, the object is constructed in memory more than once. If an object is too big for the available memory, the copies cannot be made and a timeout occurs on the client.

The revision eliminates the abstractions. It also eliminates the entire second sentence in the original passage as redundant, and it combines the fourth and fifth sentences. The topic sentence, though still general, is much easier to understand. The two sentences after it are more specific in that they state an exception and the reason for it.

In sum

Concrete information is easier to understand than abstract information. You can make information concrete through the following elements:

- Examples, samples, and scenarios
- Similes and analogies
- Specific language
- Graphics

This chapter provides guidelines to help you effectively use these elements. Refer to the examples in the chapter for practical applications of the guidelines.

When you review technical information for concreteness, you can use the checklist on page 179 in two ways:

- As a reminder of what to look for, to review the information thoroughly
- As an evaluation tool, to determine the quality of the information

Judging by the number and severity of items that you found, decide how the information ranks on each guideline for this quality characteristic. You can then add your findings to "Quality checklist" on page 387, which covers all the quality characteristics.

Although the guidelines are intended to cover all areas for this quality characteristic, you might find additional items to add to the list for a guideline.

Guidelines for concreteness	Items to look for	Quality rating
Choose examples that are appropriate for the audience and subject.	• Examples are appropriate for the audience. • Concepts have appropriate examples. • Tasks and task steps have appropriate examples. • Reference information has appropriate examples.	1 2 3 4 5
Use focused, realistic, accurate, up-to-date examples.	• Each example meets a particular need. • Examples are realistic. • Textual and code examples are accurate and up to date.	1 2 3 4 5
Make examples easy to find.	• Textual and code examples are easy to find. • Examples are in the user interface. • The start and end of examples are clear.	1 2 3 4 5
Make code examples easy to adapt.	• Code examples have clear variables. • Code examples have comments. • Large code examples are provided online.	1 2 3 4 5
Use scenarios to illustrate tasks and to provide overviews.	• Scenarios are provided where needed. • Scenarios are not abstract.	1 2 3 4 5
Set the context for examples and scenarios.	• Introductions match examples and scenarios. • Text that is introduced as an example is an example.	1 2 3 4 5
Relate unfamiliar information to familiar information.	• Analogies and similes are provided where needed. • The basis for comparison is clear.	1 2 3 4 5
Use general language appropriately.	• General language avoids abstractions. • General language is used sparingly.	1 2 3 4 5

Note: The scale for the quality rating goes from very satisfied (1) to very dissatisfied (5).

Chapter 7

Style

To me, style is just the outside of content, and content the inside of style, like the outside and inside of the human body—both go together, they can't be separated.
—Jean-Luc Godard

Style is the correctness and appropriateness of writing conventions. It is an expression of the "look and feel" of information. Developing a style for technical information means following certain conventions, standards, and rules to ensure consistency. Style also entails making choices about tone, grammar, punctuation, and presentation.

Style and content are inextricably linked. Style distinguishes a piece of writing much as accessories such as white gloves and a top hat can distinguish an outfit or occasion. However, style isn't simply decoration that you apply to information; style helps users understand the information.

Generally, editors create *style guidelines*, which are rules collected in a style guide, for the department or company to ensure consistency across sets of information. Not only does this consistency lend credibility and corporate identity to a company's publications, it also helps to deliver information that users expect from a product or line of products.

Style guides exist for many types of industries. The newspaper industry, for example, has the *AP Stylebook* and the *UPI Stylebook*. The medical industry follows the *American Medical Association Manual of Style*. Many academic and technical publishing houses follow the *Chicago Manual of Style*. These style guides provide quick answers to questions of formatting, text labels, mechanics, punctuation, tone, use of graphics, word choice, and other conventions that writers and editors must follow.

Many companies that publish technical information develop specific style guides that cover conventions in finer detail than a guide like the *Chicago Manual of Style*. Some style guides don't deal with information such as online help and reference material that isn't read sequentially. Also, a specific project might have its own detailed guidelines that cover situations in which the project style expands on or differs from the primary style guide.

Most companies also produce specific templates and boilerplate text for various types of information. You can use templates and boilerplate text to ensure that your writing has consistent appearance, organization, legal text, and other required standard information.

Some of the suggestions in this chapter overlap with aspects of task orientation, accuracy, clarity, and visual effectiveness. The style guidelines about using an active style and appropriate mood, for example, reinforce the task orientation guideline about providing clear, step-by-step instructions. The accuracy quality characteristic extends also to using correct grammar. Clarity provides the rationale for many style decisions that become style guidelines. Templates and style guidelines can help ensure visual effectiveness.

To make information stylistically appropriate and consistent, follow these guidelines:

- ❑ **Use correct grammar.**
- ❑ **Use correct and consistent spelling.**
- ❑ **Use consistent and appropriate punctuation.**
- ❑ **Write with the appropriate tone.**
- ❑ **Use an active style.**
- ❑ **Use the appropriate mood.**
- ❑ **Follow template designs and use boilerplate text.**
- ❑ **Create and follow style guidelines.**

Use correct grammar

Using correct grammar is not as simple as it sounds. In speaking, people often use grammar incorrectly, and because incorrect grammar is so prevalent, you might not always know what is right and wrong. Nevertheless, you should verify that your sentences are grammatically correct. Making grammatical errors in your information lessens your credibility and the credibility of your company.

Incorrect grammar is not only confusing for English speakers, but it can also make the translator's job more difficult. A pronoun that is used incorrectly or a dangling modifier can produce translated information that is inaccurate or nonsensical.

Resolving grammatical issues is not just about correcting errors. Sometimes you need to decide whether breaking traditional grammatical rules is appropriate for your audience and purpose. For example, some grammar books still consider splitting an infinitive or ending a sentence with a preposition to be errors. However, some phrases are awkward if you don't split the infinitive or end the sentence with a preposition.

Consider the following original and revised sentences. Both sentences tell users to check that a file links to a column in a table. However, the original sentence follows the traditional, formal rule that states that writers should not end sentences with prepositions:

Original

Check the file to which the column links.

Revision

Check the file that the column links to.

The revision uses a contemporary, informal style. Because few people speak in sentences like the original sentence, many readers might think the original sentence is awkward or pretentious. Ending the sentence with the preposition sounds natural to most readers.

The following sentences show the awkwardness of trying to avoid a split infinitive:

Original

To estimate the cost of the repairs thoroughly, we will need more inspections.

First revision

To estimate thoroughly the cost of the repairs, we will need more inspections.

Second revision

To thoroughly estimate the cost of repairs, we will need more inspections.

The original and the first revision follow the formal grammatical rule that says do not insert a word between *to* and the verb, so the first two sentences might sound awkward to most readers. The second revision splits the infinitive *to estimate*, yet it is easier for most readers to understand. Knowing your readers and their needs will help you decide how to construct your sentences.

The following grammatical problems or issues often occur in technical information:

- Sentence fragments
- Incorrect pronouns
- Dangling modifiers

Check for sentence fragments

Fragments are incomplete sentences, but using a fragment isn't always wrong. Headings, titles, captions, and some information in table cells are generally fragments. Moreover, your style guide might allow or mandate that you use fragments in some situations. For example, some style guides allow you to start a bulleted list with a fragment. However, do not write fragments within paragraphs.

Inappropriate fragments typically occur when writers misuse punctuation. Misusing the semicolon is the most common way that writers create fragments, as in the following original sentence:

Original

The client software creates specific output files; such as XML files and SQL files.

Revision

The client software creates specific output files such as XML files and SQL files.

Semicolons are similar to periods in that semicolons must be followed by a complete sentence (except for items in a list). The part of the original sentence after the semicolon is not a complete sentence because it doesn't have a verb. The revised sentence correctly connects the phrase *such as XML files and SQL files* to the rest of the sentence.

Some style guides might suggest that you use a fragment before a bulleted or numbered list, as in the following example:

> To start the Cube Editor:
>
> 1. Select a cube.
> 2. Click . . .

You can also use a full sentence to start the list:

> To start the Cube Editor, follow these steps:
>
> 1. Select a cube.
> 2. Click . . .

Check your style guide about using fragments before lists, and always correct fragments within paragraphs.

Correct pronoun problems

Two common problems with pronouns occur with two of the relative pronouns (*which* and *that*) and with indefinite pronouns.

Relative pronouns: *which* and *that*

Relative pronouns introduce dependent clauses, which can be restrictive or nonrestrictive. Unlike restrictive clauses, nonrestrictive clauses can be dropped from a sentence without changing the meaning; they simply add nonessential information. Many style guides recommend using *that* with restrictive (or essential) clauses and *which* (plus commas) with nonrestrictive clauses.

Careful readers and translators recognize the distinction between *that* and *which*. If you don't use them correctly, your readers and translators might misinterpret your information.

The following sentence shows how misusing *which* can cause confusion:

Original

> The tool kit provides a set of utilities which exploits some of the security features of InfoDBase, Version 2.

Revision

> The tool kit provides a set of utilities that exploits some of the security features of InfoDBase, Version 2.

The original sentence uses the word *which* ambiguously. A user or translator might interpret the *which* clause as nonrestrictive and infer that the tool kit provides only one set of utilities. However, the writer meant to refer to a certain set of utilities out of many sets. The writer should have used *that*, as in the revised sentence, instead of *which* to show that the information in the relative clause is restrictive.

Using the relative pronoun *which* and commas indicates that a clause is less important and that the clause doesn't restrict the meaning of the term or phrase that it modifies. Remember to use commas around nonrestrictive clauses.

Indefinite pronouns

Another common error with pronouns occurs when a pronoun does not agree in number with the noun to which it refers. To avoid creating such errors, you need to know which pronouns are singular and which are plural.

Pay close attention to indefinite pronouns such as *all, some, none, each, several, many, either,* and *neither*. These pronouns can be troublesome because they are so often used incorrectly in everyday speech. Some grammar books also disagree on the use of *none*. Some grammarians consider *none* to be singular in all situations. Others treat *none* as singular or plural. *None* can be read as *not one* or *not all*. For example, the following two sentences can be correct depending on which theory you apply to the use of *none*:

> None of the printers is online.
>
> None of the printers are online.

If you think that the use of *none* might be misinterpreted, rewrite the sentence.

Remember the following rules when you use pronouns:

- *Each, either,* and *neither* are always singular.
- *Few, both, several,* and *many* are always plural.
- *All, any, most,* and *some* can be singular or plural depending on the noun in the *of* phrase that follows them.

In the following sentence, the singular pronoun *each* must be followed by another singular pronoun:

Original

Place each of the disks in their own packages.

Revision

Place each of the disks in its own package.

The revised sentence correctly uses *its* to replace *each*. However, not all indefinite pronouns are singular. The pronoun *some* can be singular or plural. For example, both of the following sentences are correct:

Some of the **servers** are stored in their own containers. (plural subject)

Some of the **software** is stored in its own container. (singular subject)

You must check the word in the *of* phrase that directly follows *some* to know whether to use a singular or plural verb.

Correct dangling modifiers

A *dangling modifier* is a phrase or clause that cannot logically modify another element in the sentence. The sentence element that is supposed to be modified might not exist or is in the wrong grammatical form. Dangling modifiers are never appropriate in any type of writing.

Dangling modifiers are often created when writers combine an introductory participial phrase (*-ing* phrase) with a passive sentence. The following original sentence implies that the connections assembled the processor:

Original

Assembling the processor, the connections were damaged.

First revision

> Assembling the processor, we damaged the connections.

You can also avoid confusion by using a full clause instead of the *-ing* phrase:

Second revision

> When we assembled the processor, we damaged the connections.

The second revision is no longer ambiguous or confusing because it includes a subject and a verb. Readers know precisely who is performing the action.

As you write, be careful when you use participial phrases and passive sentences. Such combinations often produce dangling modifiers.

For information about misplaced modifiers and ambiguous phrases, see the clarity guideline "Place modifiers appropriately" on page 114.

Use correct and consistent spelling

Dealing with spelling can be more troublesome than correcting simple typographical errors. A word that is spelled inconsistently or incorrectly might confuse users and translators.

You will most likely need to decide how to spell new technical terms. Many industries and disciplines invent terms, some of which have more than one spelling. For example, the computer industry has created the following spellings:

- database and data base
- email, E-mail, and e-mail
- filename and file name
- meta data, metadata, and meta-data
- on line, online, and on-line
- runtime, run time, and run-time

In the following original passage, the writer spells *e-mail* inconsistently:

Original

> All e-mail files are stored in the database. You can retrieve email files if you have access to the server.

Revision

> All e-mail files are stored in the database. You can retrieve e-mail files if you have access to the server.

The revised passage spells *e-mail* consistently. Users don't have to decide if the two different spellings of *e-mail* have different meanings.

Either spelling of *e-mail* might be acceptable, depending on the style guide. However, you should choose one spelling and use it consistently.

If a dictionary does not address particular terms, consider creating a style sheet that lists the spellings that you will use for your information and when. One spelling might be appropriate for a word when it is used as a noun (such as *run time*), and another might be appropriate when the word is used as an adjective (such as *run-time*).

Before you start a writing project, consider adopting a specific dictionary so that everyone on the team uses the same spellings. In most situations, using different, nontechnical dictionaries will not create inconsistencies, especially for established terms. However, because technical terms are constantly being added to the English lexicon, you can't assume that all dictionaries adopt the same new words at the same time.

If a mainstream dictionary doesn't address a particular spelling, you can also check industry-specific dictionaries such as *Webster's New World Dictionary of Computer Terms*. However, if you don't find the term in a dictionary, determine the most accepted spelling by checking sources in the relevant industry. See the clarity guideline "Use technical terms only if they are necessary and appropriate" on page 141 for more information about using technical terms.

In addition to new spellings, check the spelling of often confused words, such as *principal* and *principle*. In the following sentence, the word *affect* is used incorrectly:

Original

XML programming has a significant affect on information storage.

Revision

XML programming has a significant effect on information storage.

The revised sentence uses the correct term, *effect*. Most spelling checkers cannot determine whether words such as *affect* or *effect* are used correctly in a particular sentence. They can only flag it for the writer to check.

Consider the following often confused words. Make sure that you use them correctly:

- accept and except
- access and assess
- advice and advise
- affect and effect
- capital and capitol
- ensure and insure
- its and it's
- principal and principle
- their, they're, and there
- to, too, and two
- whose and who's
- your and you're

See the accuracy guideline "Use tools that automate checking for accuracy" on page 67 for more information about finding both spelling and grammar errors.

Use consistent and appropriate punctuation

Most industries have adopted guidelines about how to use punctuation because no single authority for English punctuation exists that provides rules for every situation. Moreover, just as writing styles change over time, some of the guidelines that govern the use of punctuation also change. For example, the newspaper industry currently prefers not to use a serial comma before a coordinating conjunction, whereas in technical writing this comma is often mandatory.

Not all style guides agree on the use of punctuation. For example, non-American style guides typically say to use single quotation marks around new terms. As another example, many technical style guides suggest that you avoid using a pair of dashes to set off parenthetical information. In the following sentence, notice that the information surrounded by dashes is not important or restrictive:

Original

> Additions to the project schedule—such as deadlines, requirements, and project codes—must be entered before the end of the month.

Revision

> Additions to the project schedule (such as deadlines, requirements, and project codes) must be entered before the end of the month.

The dashes suggest that the information is important, and users or translators might not be able to determine if the information inside the dashes is restrictive or nonrestrictive.

Using the dash also creates an informal tone. In most cases, you can replace dashes with parentheses or commas if the information is nonrestrictive.

Combinations of punctuation can also be confusing. The most common combination is commas or periods with quotation marks. Both of the following sentences are correct, depending on the style guide that you adopt:

> For more information, see "Adding dimensions to cubes."
>
> For more information, see "Adding dimensions to cubes".

Before you start a writing project, consider how you will apply punctuation and mechanics. In addition to colons, commas, and semicolons, be especially careful about using the following punctuation:

- ❑ Apostrophes
- ❑ Brackets
- ❑ Em dashes and en dashes
- ❑ Hyphens
- ❑ Parentheses
- ❑ Slashes
- ❑ Single and double quotation marks (especially when they are used with commas and periods)

You can also consider how to use other mechanical elements such as numbers, symbols, and units of measurement. For example, should you spell out numbers or use numerals? How would you write mathematical expressions? After a number, should you spell out the unit of measure (such as megabyte) or use the abbreviation (such as MB)? And if you use an abbreviation, should it be preceded by a space or not?

Write with the appropriate tone

Tone conveys how a piece of writing "sounds" or how it feels to the user. Tone expresses the writer's attitude toward the subject and the audience. An appropriate and consistent tone helps the user do the task, learn the concepts, or make the right decisions without needing to interpret your rhetoric.

The tone of technical information must be helpful: writing concrete and accurate information establishes a helpful tone. The tone must be direct: writing clear information establishes a direct tone. The tone must also be authoritative: writing task-oriented, accurate, and complete information establishes an authoritative tone.

Many tones are possible in technical writing. To achieve the tone that works best for a particular situation, you must understand the audience and the purpose for the information. For example, if you're writing a marketing brochure, you probably want to use a different tone than you would for a reference manual. As another example, a popular line of how-to books addresses novice users who want the basics foremost and details only at an easy-to-grasp level. The authors of this series of books write in a very friendly, humorous tone.

However, if you produce information that will be translated, avoid humor and sarcasm because they rely on specific cultural references and do not translate easily from one language to another. Be sure that you understand your users' needs before you use humor, sarcasm, or metaphors in your information.

In the following sentence, the writer is trying to be friendly while providing a requirement:

Original

Of course, you might not be able to print if you don't first configure your printer.

Revision

Before you print, you must configure your printer.

In the original sentence, the writer tries to be polite and friendly. However, this tone makes the sentence wordy and unclear. The revised sentence is more direct and clearer than the original sentence.

Sometimes, writers use colloquial or idiomatic expressions to create a friendly, less intimidating tone. However, as with humor, colloquial and idiomatic expressions are difficult or impossible to translate accurately. Such expressions are also difficult for nonnative English speakers to interpret. To help translators create unambiguous translations, use a straightforward, noncolloquial style. Avoid idioms and slang, which nearly always cause problems for translators. In addition, be sure to write in such a way that the tone does not alienate nonnative speakers of English. Consider the use of slang and idioms in the original sentence:

Original

> This section describes a whole slew of geeky code conventions used throughout this book, so hang on.

Revision

> This section describes the code conventions that are used throughout this book.

Although the original sentence tries to make novice users feel less intimidated by the technical jargon, the humor relies on the knowledge of a particular subsection of American usage. Words and phrases such as *whole slew*, *geeky*, and *hang on* cannot be easily translated in other languages. The informal, somewhat flippant tone might even be offensive in some cultures. Because the revision uses direct, formal English, it is less likely to confuse or offend users.

When you write for the Web, depending on the type of information that you provide, you can often use a tone that is engaging and friendly. A word of caution, however: People all over the world access the Web, so you should write for an international audience. Your Web pages can be lively and entertaining, while maintaining a tone that is professional and considerate of nonnative speakers of English.

The following sentence uses an idiomatic expression unnecessarily.

Original

> InfoTool is bending over backward to be accommodating.

Although *bending over backward* is an idiom that most native English speakers have heard and understand, it is difficult to picture it applied to a software tool.

Revision

> InfoTool now offers several options that provide the flexibility that you need.

The revised sentence expresses the same message as the original sentence in a style that is descriptive without being baffling. See the combined quality guideline "Applying quality characteristics to information for an international audience" on page 344 for more information about writing for international audiences.

Technical information is sometimes written with a pretentious tone. To be seen as credible and authoritative, some writers use complex sentences, abstract technical terms, and a wordy style. This tone can be confusing and intimidating. Consider the tone in the following sentence:

Original

> One must recognize that regional expenditures could be reduced by the new accounting measures.

First revision

> Everyone must know that the new accounting measures can reduce regional expenses.

The original sentence is not only confusing and wordy, but it is also pretentious. Using the pronoun *one*, passive voice, and long verb phrases gives the sentence a pretentious tone. Such a tone tends to alienate readers rather than engage them.

If the idea that *everyone must know* is not important, you can omit it from the sentence:

Second revision

> The new accounting measures can reduce regional expenses.

Use the following guidelines when you write so that your tone is appropriate and consistent for a technical audience:

- ❑ Write for an international audience.
- ❑ Be helpful, concrete, and accurate.
- ❑ Be direct and clear.
- ❑ Maintain an authoritative tone without being pretentious.

Use an active style

Writing directly to users is the best way to convey information. Unnecessary passive sentences and complicated verb tenses can obscure the meaning of your information.

To help you create an active style for your information, follow these guidelines:

- Use active voice whenever appropriate.
- Write with the present tense.

Use active voice

Active voice puts the agent or performer of an action at the beginning of a sentence. If you want to emphasize who or what performs some action, use active voice. Use passive voice if you don't want to emphasize who or what performs the action. The following sentence is written in both active and passive forms:

Active: The administrator maintains the database catalogs.

Passive: The database catalogs are maintained by the administrator.

Passive voice requires a longer verb phrase and places the agent at the end of the sentence or clause, or it doesn't mention the agent at all. In the previous passive sentence, the verb is *are maintained*, which is longer than the active verb *maintains*. The agent or performer of the action is *administrator*, which comes at the end of the sentence.

Passive voice is often difficult to understand because users might not understand who is doing the action. Consider the following sentence:

Original

Up to 256 characters can be typed in this entry field.

Revision

You can type up to 256 characters in this entry field.

The message in the original sentence is unclear because users might be confused about whether the sentence means that they can add up to 256 characters in the entry field or that something else, such as the program, adds characters to the field. The revised sentence clearly states that the user can type up to 256 characters.

Sometimes passive voice is appropriate. If the action or the receiver of the action is more important than the agent, use passive voice for the proper emphasis. The following original sentence uses active voice inappropriately:

Original

> The system saves the file to the database.

Revision

> The file is saved to the database.

The original sentence emphasizes an action (*saves*) and an agent of the action (*the system*). The important part of the sentence is that the file gets saved; the idea that the system saves the file doesn't matter to the user.

Be careful with passive sentences in which the agent of the action is ambiguous and users must know who the agent is. The following original sentence doesn't tell users who or what performs the action:

Original

> The metadata is exported to the new catalog tables.

Does the user export the metadata, or does the system or program perform this task? If the writer meant that the user should export the metadata, the sentence should use the imperative mood, as in the first revision, or the indicative mood with second person, as in the second revision:

First revision

> Export the metadata to the new catalog tables.

Second revision

> You can export the metadata to the new catalog tables.

If the system or program exports the metadata and the user must understand that information, the sentence should be rewritten accordingly:

Third revision

> The export utility exports the metadata to the new catalog tables.

In all of the revised sentences, the user knows exactly who or what performs the action.

Use the present tense

Present tense verbs help you keep your writing style active and direct. You should use the present tense for your verbs whenever possible. The present tense is easier to translate and more engaging. Use the past or future tense only when the present tense would be misleading or illogical.

In the following original sentence, the writer uses the past tense unnecessarily:

Original

The Web pages were uploaded during the installation.

Revision

The Web pages are uploaded during the installation.

Although both sentences use passive voice, the original sentence uses the past tense unnecessarily. (The passive voice is appropriate because users don't upload the Web pages. The installation program does this automatically.) When the pages are uploaded is not important. The revised sentence uses the present tense appropriately.

Unless you have a good reason to use the past or future tense, always write in the present tense.

Use the appropriate mood

Mood describes the communicator's attitude toward or relation to the action. Verbs can express the indicative, imperative, or subjunctive mood. Writers of technical information use mainly the indicative mood for conceptual and reference information and the imperative mood for task information. The subjunctive mood, which expresses wishes, doubts, and desires, is rarely appropriate. A subjunctive statement can confuse users and can be difficult to translate into languages that do not have the equivalent verb quality.

The indicative mood is used to state facts and opinions, whereas the imperative mood expresses a command or request. With its implied *you*, the imperative mood addresses the user and clearly identifies the action that the user should perform. The following sentence is from a procedure:

Original

The application can be started.

First revision

You need to start the application.

Second revision

Start the application.

The original sentence does not tell the user to perform an action. It is also written in the passive voice, which makes the sentence more confusing. The first revision is an improvement because it brings the user into the picture by using the second person and the indicative mood, but this revision simply describes an action; it doesn't tell the user to do anything. The second revision shows the instruction as it should be: in the imperative mood.

Use the indicative mood to state facts and the imperative mood to instruct the user. For more information about using the imperative mood in task information, see the task orientation guideline "Provide clear, step-by-step instructions" on page 36.

Follow template designs and use boilerplate text

Using the same template across sets of information, such as for related products, ensures that the information maintains a consistent structure and appearance. Creating a template for every new project is a waste of effort.

Using boilerplate text ensures that your legal notices or other standard information is accurate and complete.

To save time and to prevent possible inaccuracies in your information, follow these guidelines:

- ❑ Create and reuse templates.
- ❑ Use boilerplate text to ensure inclusion of necessary information.

Create and reuse templates

A *template* is a collection of styles or tags that define the structure and appearance of a document. Some templates offer only a minimal number of styles, so writers, editors, and visual designers must define styles as needed. Other templates are very rigid and don't allow writers to redefine items such as fonts, tabs, character sizes, or line spacing.

Minimally defined templates might allow writers more flexibility to present information creatively, but writers run the risk of delivering information that is inconsistently presented, which could be confusing and distracting to readers. However, even though rigid templates keep writers from changing styles to fit particular needs, these templates also ensure consistent presentation. If you are involved in creating templates for your information, carefully consider the styles that you adopt and decide whether writers should be allowed to deviate from those styles.

This book uses a specific set of styles. For example, the styles define the fonts and sizes for each heading and line of body text, the line spacing, the width of the margins, and the appearance of tables. Using a template allows writers to concentrate on content instead of worrying about layout.

For more information about maintaining structural consistency throughout your information, see the completeness guideline "Use patterns of information to ensure proper coverage" on page 90 and the visual effectiveness guideline "Use visual elements logically and consistently" on page 295.

Use boilerplate text to ensure inclusion of necessary information

Boilerplate text is standard text that can be inserted into documents with few or no changes. Most companies use boilerplate text for legal and copyright statements, trademarks, or warranty clauses. If your information will be published outside your company, you will probably need preapproved legal text, notices, and contact information. Not adding the correct boilerplate text can create serious legal problems for your company.

The following passage is a portion of legal boilerplate text from an InfoDBase information unit:

> INFODBASE CORPORATION PROVIDES THIS PUBLICATION AS IS WITHOUT WARRANTY, EITHER EXPRESS OR IMPLIED, INCLUDING, BUT NOT LIMITED TO, THE IMPLIED WARRANTIES OF NON-INFRINGEMENT, MERCHANTABILITY, OR FITNESS FOR A PARTICULAR PURPOSE. Some states do not allow disclaimer of express or implied warranties in certain transactions. Therefore, this statement might not apply to you.
>
> This information can include technical inaccuracies or typographical errors. Changes are periodically made to the information herein; these changes will be incorporated in new editions of the publication. InfoDBase Corporation might change the product or products or the program or programs that are described in this publication at any time without notice.

Legal boilerplate text is generally complicated information that is provided to protect the company from litigation. You are rarely allowed to change or remove such legal information. Technical editors or lead writers should ensure that boilerplate text is current and appropriate for specific products or services.

In addition to boilerplate text, most industries mandate that you use warning or caution labels only for specific situations. Some companies provide boilerplate text for danger, caution, warning, or attention labels. For example, most computer software style guides say to use *caution* labels only when an action could physically harm the user or others.

The following passage uses text labels inappropriately. The style guide for the InfoDBase product requires that caution notices be used only for situations that are hazardous to people and not to insert an exclamation point. The style guide also specifies that attention labels be used to indicate that an action might cause loss of property or data.

Original

> **Caution!** Do not attach cable A to the device until the red indicator light changes to green unless you want to reconfigure your settings.
>
> . . .
>
> ***Warning*:** If you delete the record from the table by clicking **Delete**, the record will be deleted, and the program will close.

Revision

> **Caution:** Do not touch the wires. They might be hot.
>
> . . .
>
> **Attention:** If you delete the record from the table by clicking **Delete**, the record will be deleted, and the program will close.

In the original sentences, the writer uses different words inconsistently to describe potentially serious information. The revision uses the correct label, text, and highlighting as specified by the product's style guide.

Before you write, consult your style guide about when to use text labels, highlighting, and boilerplate text to warn users, make recommendations, or provide notes. Use boilerplate text and text labels when appropriate. Do not interchange text labels or highlight them differently from one place to the next.

Create and follow style guidelines

Style guidelines, whether they come from a style guide or a style sheet, help you and your team consistently apply specific conventions to your information. For example, style guidelines can specify what text to highlight, what tone to use, or what spellings to use. A style guide might contain hundreds of entries that describe how to deal with everything from apostrophes to Web addresses.

Information presentation that is not uniform is distracting and potentially confusing for readers. Style guidelines help ensure that writers working on related information do not introduce variations that will create inconsistent presentation. Before a writing project begins, your team should create or adopt style guidelines.

The easiest way to prepare a set of style guidelines is to adopt a published style guide. If necessary, the lead editors and writers can adapt specific guidelines in the published style guide for their company or projects.

To help everyone on your project follow style guidelines, you can create quick-reference *style sheets*. A style sheet is generally shorter than a style guide, and it provides a list of project-specific style guidelines. In addition to expansions on the style guide, a style sheet can also document deviations from the style guide.

For example, your style guide might say to keep your tone formal, direct, and neutral. However, you need to write information that will be included in marketing materials, so you want to provide lively, less technical information. Therefore, your style sheet for marketing information might say to use contractions, metaphors, and even humor.

Figure 14 on page 204 shows the relationship between style guides and style sheets. Externally published style guides typically provide the foundation for internal or project-specific style guides and style sheets.

Figure 14. Relationship of style guides to style sheets.

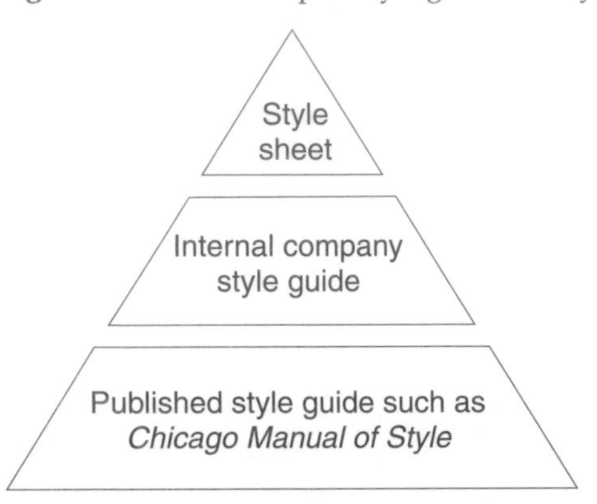

Style sheets and style guides can be arranged alphabetically, by style category, or any other logical arrangement. The more writers and editors can do before actual writing begins to ensure consistent style across information, the better the result will be, and the less copy editing and updating that will be required.

If you need to decide what specific conventions to use for a set of style guidelines, such as whether to allow *and/or*, whether to capitalize *Web*, or whether to use *mice* as the plural of *mouse* (the computer device), consider the following advice:

1. Consult relevant authoritative sources for similar situations. Check more than one source to see whether authorities disagree on the subject.
2. If you find no obvious similar situations, isolate the underlying issue. Look for precedents that apply to the issue or for guidelines that define style conventions in the general area.
3. Consider the consequences for users, particularly translators and non-native speakers of English. Are users likely to benefit from this style decision? Could they be confused by it and misinterpret the meaning?
4. Consider the ease of carrying out possible style decisions. How easy would each be to apply, probably by people other than just yourself? Is the reasoning straightforward, or does it have some exceptions?

Whatever your source for style guidelines, if you follow them consistently, you'll provide your users with a reliable presentation that won't interfere with their ability to use the information easily.

In addition to checking and deciding on conventions of grammar, punctuation, spelling, and tone, consider the following guidelines:

- ❑ Provide practical and consistent highlighting.
- ❑ Present list items consistently.
- ❑ Use unbiased language.

Provide practical and consistent highlighting

When you use highlighting consistently and predictably, users can pay attention to the content. They don't need to stop and think about why a certain term is in boldface and another is in italics, or what the difference is between entering a command that is displayed in lowercase and one that is displayed in all capital letters.

Assume that each of the following instructions is from different parts of a product's information:

Original

> Press "ENTER."
> . . .
> Enter the `Search` command
> . . .
> At a command prompt, type `help` and press *Enter.*

The problems are obvious when you look at the original instructions together. The word *enter* appears in three different styles, two of which apply to the same usage. The second and third instructions use the same highlighting (monospace) for different situations. The writer uses different types of highlighting (all capital letters plus quotation marks versus a capital letter plus italics) to refer to a key. Even if the writer provides an explanation of highlighting, these discrepancies are confusing and might cause users to make mistakes.

Revision

> Press Enter.
> . . .
> Enter the SEARCH command.
> . . .
> At a command prompt, type `help` and press Enter.

The revised instructions reduce the amount of highlighting and do not use conflicting styles. Users who quickly scan the instructions will not stumble over the confusing highlighting in the original text.

The sample style sheet in Figure 15 describes how to apply different types of highlighting for the InfoDBase product information:

Figure 15. Style sheet for InfoDBase: Highlighting.

Element	**Highlighting**	**Example**
API calls	none	sqlurlog
Book and CD-ROM titles	italics	See the *InfoDBase Reference* . . .
Code examples	monospace	Type the following command: `xclient userid pwd import`
Commands	bold	The **exit** command is not available.
Control names and check boxes	bold	Use the **New file** radio button . . .
Directory names	monospace	The XML file is in the `c:\temp` directory.
Entered data	monospace	Type `exit` and press Enter.
Entry fields	bold	In the **Table name** field, type . . .
Example text	monospace	For example, `access denied` indicates that . . .
File names	monospace	Open the `readme.txt` file.
Function names	none, all caps	The DIGITS function processes . . .
Interface objects	bold	Click the **System menu** icon.
Keys	none	Press Alt+F1
Menu names and choices	bold	To create a new project, click **Project —› New**.
Message text	monospace	`Action complete` message . . .
New terms (where defined)	italics	A *hierarchy* is a . . .
Parameter variables	italics	Set *userid* to your last name . . .
Window and wizard names	none, initial caps	In the Project window . . . The Create Dimension wizard . . .

Present list items consistently

Guidelines for list items vary among style guides. However, many technical style guides recommend the following guidelines for lists:

- Use a complete sentence for the list lead-in sentence unless your style guide specifies that you should use fragments.
- Include at least two items in a list.
- Make list items parallel.
- Start each list item with a capital letter.
- If one list item ends with a period (because it is a sentence), use periods on all items.
- If none of the list items are sentences, don't use periods.

The following list violates several of the previous guidelines:

Original

The metadata object catalog can contain:

- Dimensions
- hierarchies that can be from any dimension
- cubes
- facts.

The original list does not use a complete sentence for its lead-in sentence, it does not use capitalization and punctuation consistently, and it is not parallel:

Revision

The metadata object catalog can contain the following objects:

- Dimensions
- Hierarchies (from any dimension)
- Cubes
- Facts

In the revision, the introduction is a complete sentence, each list item starts with a capital letter, and no list item uses a period.

For more information about presenting lists, see the clarity guideline "Keep lists short" on page 129.

Use unbiased language

As more and more products are marketed globally, you should pay attention to biased language in your information. Avoid making assumptions about gender or race in your writing.

For example, suppose that a scenario includes a list of fictitious names, phone numbers, and e-mail addresses. The following table includes only male Anglo-Saxon names:

Original

Last name	First name	Phone number	E-mail address
Carter	Adam	(908) 954-6833	carter@domain.com
LaBlanc	Paul	(301) 266-2110	plablanc@oursite.net
Taylor	John	(210) 454-3241	taylorj@thissite.com
Smith	Joseph	(408) 555-4444	smith@site.com

Revision

Last name	First name	Phone number	E-mail address
Carter	Andrea	(908) 954-6833	cartera@domain.com
Jiminez	Carlos	(301) 266-2110	jiminez@oursite.net
Nguyen	Thinh	(210) 454-3241	tnguyen@thissite.com
Bartelli	Maria	(408) 555-4444	mabartelli@site.com

Even though the names in the table don't refer to real people, the revised names to some extent reflect the diversity of an international audience and avoid the sexism of an all-male list.

Include references to gender or race only if they are relevant. The following sentence is acceptable in scientific information:

> The clinical study included 10 female patients and 12 male patients.

Also, be careful with references to occupations. Don't write a sentence so that you imply that all nurses or administrative assistants are women, or that all doctors and programmers are men. The following sentence makes an assumption about the gender of an administrative assistant:

Original

> An administrative assistant helps her manager with everything from data processing to scheduling meetings.

You can revise the original sentence by changing the nouns and pronouns to plural:

Revision

> Most administrative assistants help their managers with everything from data processing to scheduling meetings.

When you correct sexist language, avoid using *he or she* as in the following original sentence:

Original

> If the programmer keeps an accurate account of his or her programming time, he or she can determine his productivity.

Instead of using *he or she*, use plurals, the imperative mood, or second person to restructure the sentence as in the following revised sentences:

Revision using plural

> If programmers keep an accurate account of their programming time, they can determine their productivity.

Revision using imperative

> Keep an accurate account of your programming time so that you can determine your productivity.

Revision using second person

> If you keep an accurate account of your programming time, you can determine your productivity.

Or you can write the sentence as an example by using a fictitious person's name:

Revision using a person's name

> For example, if Susan keeps an accurate account of her programming time, she can determine her productivity.

Depending on the context of a particular sentence, you can use one of the styles shown in the revisions to avoid gender-specific pronouns.

When you revise information such as this, ensure that you don't pair a singular noun with a plural pronoun:

> If a programmer keeps an accurate account of their programming time, . . .

Although many people speak this way, this sentence is grammatically incorrect because the plural pronoun doesn't agree in number with its singular noun. See "Correct pronoun problems" on page 185 for more information about how to correct pronoun problems.

In sum

A consistent and appropriate style helps users understand information. Effective style is transparent and doesn't hinder users' understanding of the information. Users notice style only when highlighting, punctuation, or spelling is used inconsistently; when sentences are ungrammatical; when the tone is inappropriate; or when legal notices are incorrect. When users pay attention to the style, they probably do not pay attention to the meaning of the information.

Use the guidelines in this chapter to ensure that technical information has the correct and appropriate style for your audience. See the examples in the chapter for practical applications of these guidelines.

When you review technical information for correctness and appropriateness of style, you can use the checklist on page 212 in two ways:

- ❑ As a reminder of what to look for, to ensure a thorough review
- ❑ As an evaluation tool, to determine the quality of the information

You can apply the quality rating in the third column of the checklist to the guideline as a whole. Judging by the number and severity of items you found, decide how the information rates on each guideline for this quality characteristic. You can then add your findings to "Quality checklist" on page 387, which covers all the quality characteristics.

Although the guidelines are intended to cover all areas for this quality characteristic, you might find additional items to add to the list for a guideline.

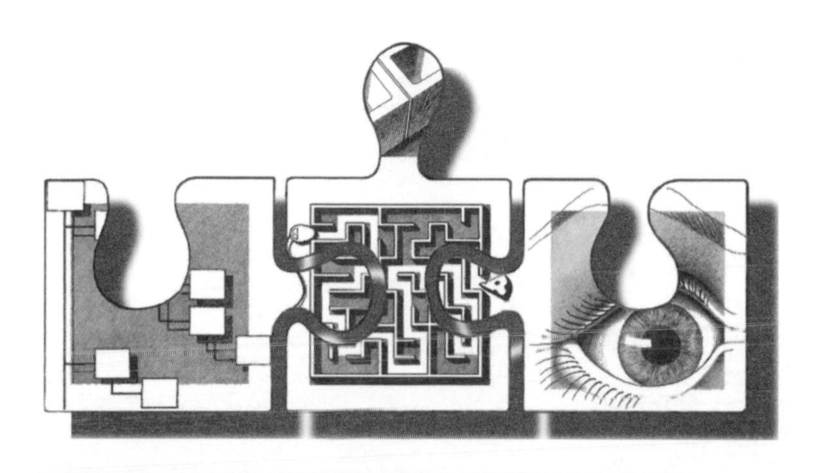

Part 3

Easy to find

Finding information is essential, whether you're a new user trying to figure out where to start and what to do or an experienced user looking for a parameter for a command. The organization, retrievability, and visual effectiveness of information contribute most to whether users can find the information that they need.

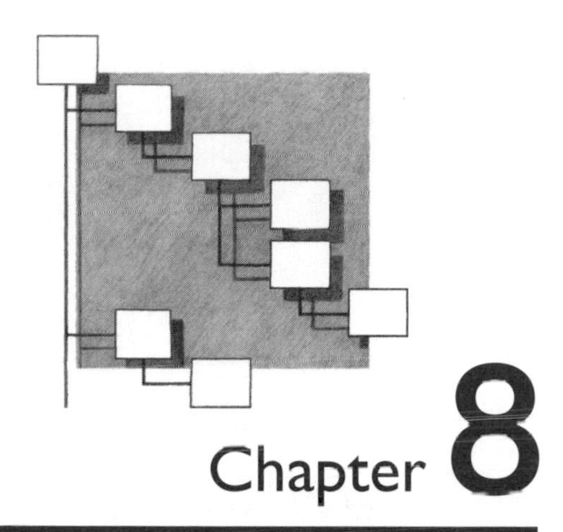

Chapter 8

Organization

Without order there is only chaos, and nothing constructive can evolve from chaos except order.
—Duncan Head

Organization is how pieces of information fit together to form a coherent arrangement of parts that makes sense to the user.

Organization applies both to the arrangement of information within a topic and to the arrangement of two or more topics (how topics fit together). You can build up larger structures by fitting together the parts. For example, you build paragraphs with sentences; you build topics with paragraphs, lists, tables, and graphics; and you build information centers, Web sites, help systems, and books with topics.

Figure 16 shows how you organize information within topics and then organize those topics to build a set of information.

Figure 16. Topics contain paragraphs, lists, and other elements; information centers, books, and online help contain topics.

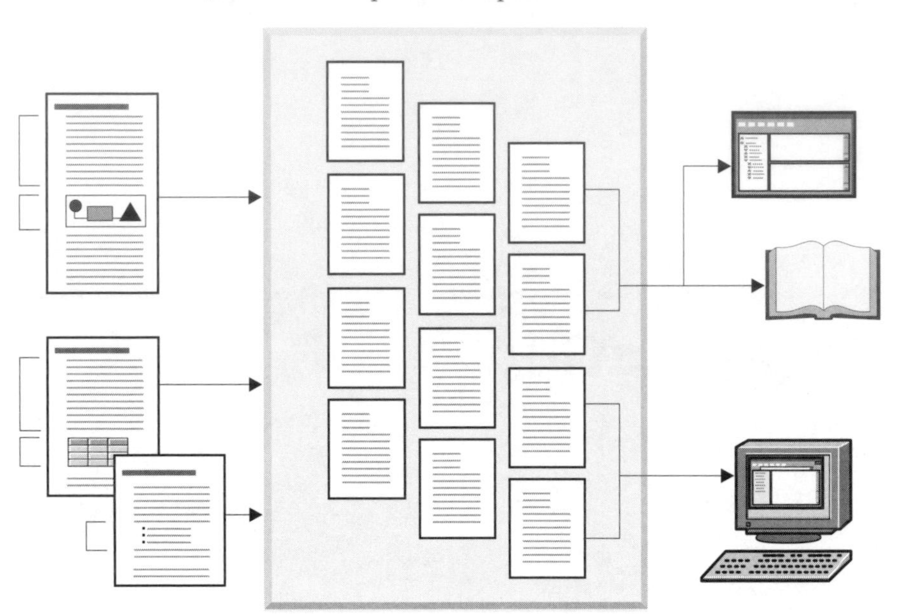

Good organization can accommodate change easily. For example, if you create a separate topic for each task, you can easily add, remove, or rearrange tasks when the product changes or when you want to provide another organization scheme for a different set of users.

Poorly organized topics are often littered with paragraphs that are marked with a note label because the pattern of the information is too rigid to accommodate change. Poorly organized navigation elements (such as tables of contents or navigation panes) can hide topics that don't fit well into the existing structure.

Well-organized information is easy to find, especially when the organization is obvious to the user. Well-organized information is also easier for writers to reuse effectively for different types of output. Organizing information properly might take more time, but doing so will better serve the user and you in the long run.

To organize information well, follow these guidelines:

- ❑ **Organize information into discrete topics by type.**
- ❑ **Organize tasks by order of use.**
- ❑ **Organize topics for quick retrieval.**
- ❑ **Separate contextual information from other types of information.**
- ❑ **Organize information consistently.**
- ❑ **Provide an appropriate number of subentries for each branch.**
- ❑ **Emphasize main points; subordinate secondary points.**
- ❑ **Reveal how the pieces fit together.**

Organize information into discrete topics by type

Topics are small, discrete, meaningful pieces of information, each with a heading. A topic, also known as a chunk, article, or module, must be long enough to cover a subject adequately and make sense if read by itself. Organizing your information into discrete topics gives you the flexibility to reuse the information or any portion of it for a variety of presentations. In addition, topics are generally short and therefore more searchable and more likely to provide users with just the information that they need.

Most technical information falls into one of three types: concept, task, or reference:

Concept

Provides background information that users might need to know before they can work with the product. For example, a topic called "Data consolidation" is a concept topic and might contain several paragraphs and a graphic that explain what data consolidation is.

Task

Provides procedural details, such as step-by-step instructions, along with information such as rationale, prerequisites, and examples. For example, a topic called "Changing authorizations" is a task topic and might contain a brief overview of changing authorizations, a list of prerequisites, steps to change the authorizations, and a list of any follow-on (postrequisite) tasks.

Reference

Provides quick access to facts that users might look up, such as conversion tables, syntax rules, message explanations, and code samples. For example, a topic called "Environment variables" is a reference topic and might contain a table that lists the operating systems, valid settings, and a default setting for each environment variable.

If you organize information into topics according to type, you can more easily keep your topics discrete, and you can provide a consistent structure so that users know what to expect. For example, if all of your task topics contain a brief overview, prerequisites, steps, and postrequisites, users know what to expect in a task topic. If you add conceptual information to the steps, users who are looking for concepts will have a harder time finding that information, and users who are trying to do a task are interrupted by conceptual information. Similarly, users who are looking for syntax and examples of commands will have trouble locating the reference information if it is combined with concepts or tasks.

The following topic contains task, concept, and reference information:

Original

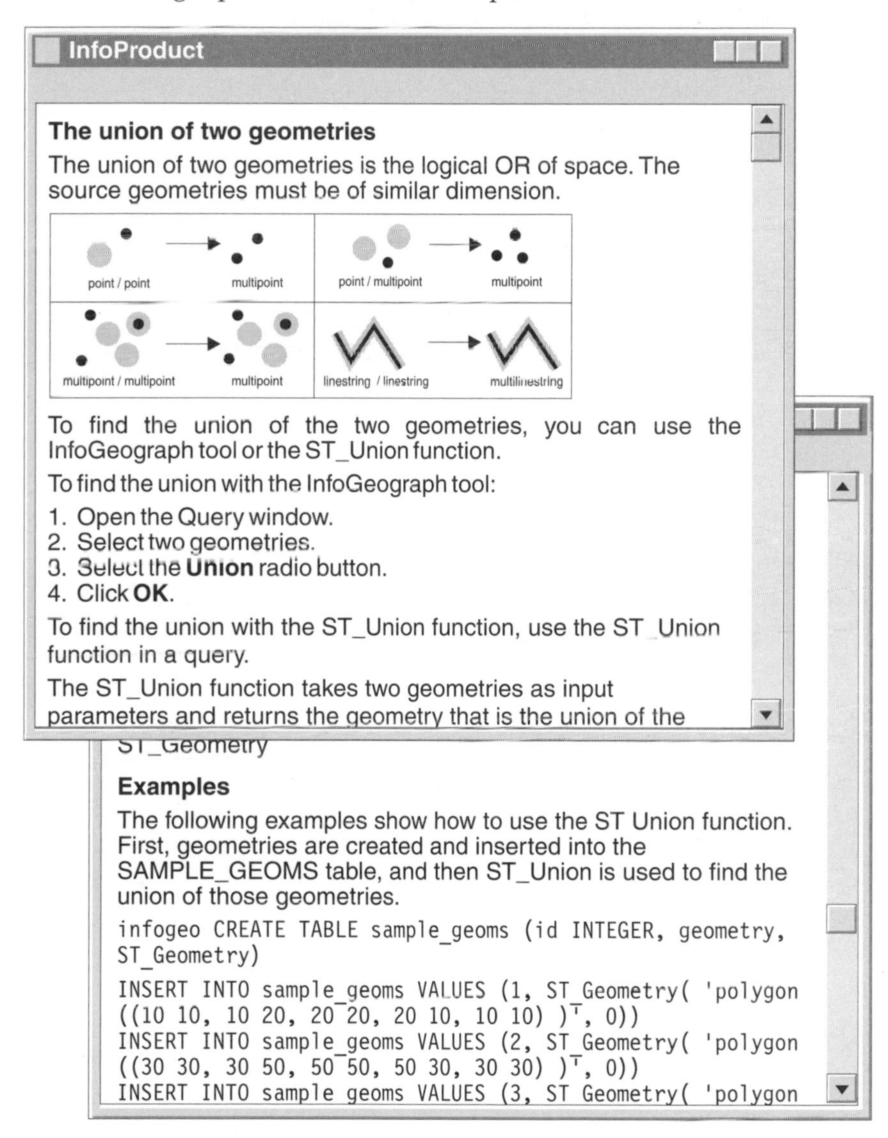

The original topic contains too much information. Users will be more likely to find the information that they want if you separate the topic into a concept topic, a reference topic, and a task topic. The concept topic can have the heading "The union of two geometries," the reference topic can have the heading "ST_Union function," and the task topic can have the heading "Finding the union of two geometries."

Revision

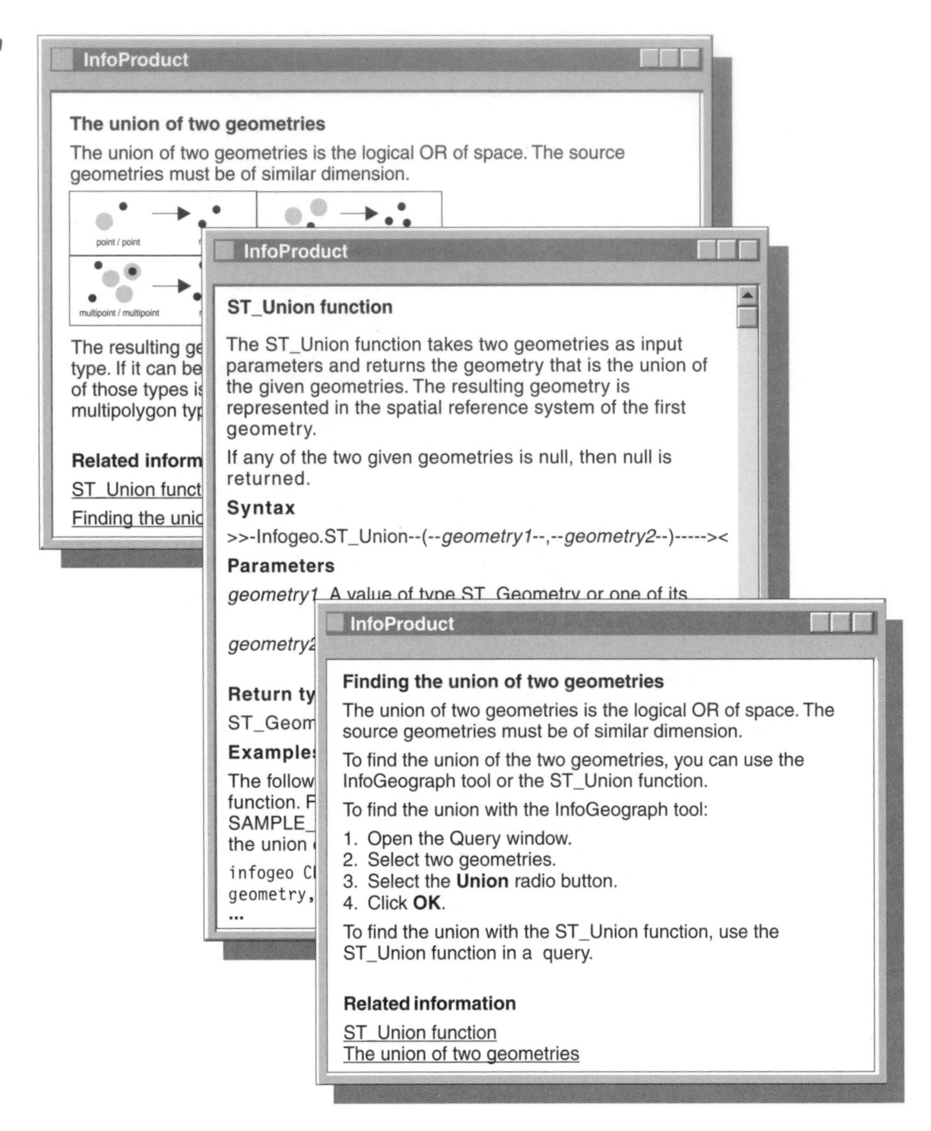

The revision shows three topics: a concept topic that explains what a union of geometries is; a reference topic that explains the syntax for the ST_Union function; and a task topic that explains how to find the union of two geometries. Each of the three topics provides links to the other two topics.

By splitting information into discrete topics that do not presuppose a specific structure or linear arrangement, you allow the topics to be organized or reused in different ways. A single topic can appear in multiple views of an information center, in a PDF file, in a printed book, and on a Web site. You can also share topics across sets of information for:

- Multiple output media
- Multiple product features
- Multiple operating systems
- Multiple audiences

In addition, you can add metadata to topics that allows users to dynamically build their own sets of information. For example, you can add metadata to a topic that specifies the operating system, type of user, product feature, and other facts about the topic. Users can build sets of information that fit their requirements based on the metadata that you specify.

Organize tasks by order of use

Because users generally complete tasks in a certain order, the information should be arranged to reflect that order. If what users do in task Y depends on what they have done in task X, the topics should be arranged so that task X precedes task Y.

The following outline shows an organization that doesn't correspond to the actual order of tasks:

Original

Getting started
- Planning for installation
- Verifying that prerequisite software is installed
- Installing InfoProduct
- Removing InfoProduct
- Testing connections
- Configuring InfoProduct

The original outline shows that the task of configuring the product follows the task of testing the connections. In addition, the original outline lists the task of removing InfoProduct as a getting started task. Users are not likely to remove a product right after they install it, nor are they likely to look for the task of removing the product under the getting started branch.

Revision

Getting started
- Planning for installation
- Verifying that prerequisite software is installed
- Installing InfoProduct
- Configuring InfoProduct
- Testing connections

Common sense suggests that users will have better luck testing their connections after they configure the product, as shown in the revised outline. The revised outline also does not include the removing task with the other getting started tasks.

Organizing tasks logically is important not only at the outline level, but also at the level of sentences and steps within each task topic. If the information in a task topic is not presented in the order that users need it, users waste time and become frustrated.

The following task topic shows a task in which the information and steps are not presented in the order that users need to know them:

Original

Installing InfoTax

To install the InfoTax software:

1. Run the *setup.exe* program. Follow the instructions that are provided on the installation screens. To access the *setup.exe* program, unzip the InfoTax software files by using a file decompression utility.
2. Before you install the software, ensure that you have enough storage space on your hard disk. The program automatically assesses your storage space, and the installation stops if there is not enough space. You will need 120 MB of free space.
3. Read the *readme* file to find out whether any differences apply to the installation for the operating system that you are using. The installation task is system dependent.
4. Create a directory in which to install the product. The name of the directory must begin with the characters *tax*.

System requirements:

- ❑ 2.0 GHz processor
- ❑ 256 MB RAM

Have your 16-digit product registration key available. You can find the key on the inside of the back cover of the *InfoTax User's Guide*.

Note: Shut down all other programs before you install InfoTax.

In the original task topic, information that users need before they install the InfoTax software is buried within the steps and is not in the order that the users need to use it. Users might start the installation task and waste time before realizing that they need a faster processor or more storage space. In addition, some key steps, such as shutting down all other programs before installing InfoTax, are in the system requirements section after the installation steps.

Revision

Installing InfoTax

Before you install InfoTax:

- ❑ Have your 16-digit product registration key available. You can find the key on the inside of the back cover of the *InfoTax User's Guide.*
- ❑ Ensure that your system has:
  - 2.0 GHz processor
  - 120 MB of free storage space
  - 256 MB of RAM
- ❑ Read the *readme* file to find out whether any differences apply to the installation for the operating system that you are using. The installation task is system dependent.
- ❑ Create a directory in which to install the product. The name of the directory must begin with the characters *tax*.

To install the InfoTax software:

1. Stop all programs that are currently running.
2. Unzip the program files by using a file decompression utility.
3. Run the *setup.exe* program. Follow the instructions that are provided on the installation screens.

In the revised task topic, all of the tasks that must be done before the installation are explained before the installation steps. The tasks in this section need not be completed in an exact order as the original task topic implies, so they are shown in an unordered list in the revision. As a result, the installation task is streamlined, with the steps in the order that the user must do them to avoid making mistakes or losing work.

See the task orientation guideline "Provide clear, step-by-step instructions" on page 36 for more information about organizing steps.

Organize topics for quick retrieval

In addition to organizing your main task topics by order of use, you need to fit concept and reference topics into your structure in a way that makes sense to the user. If you have concept and reference topics that are associated with particular tasks, you might arrange these topics with the tasks. Some concept and reference topics do not presuppose a particular task or do not fit well into a task structure. In these cases, you can arrange the topics for quick retrieval.

Concept and reference topics are often arranged either alphabetically or numerically. With an alphabetical arrangement, whether in a book or in an information center, a user who is looking at the topic "Infolist command" knows to go forward to get to the topic "Store command" and backward to get to "Grant command." Similarly, with a numerical arrangement, a user who is reading about message 120 knows to go forward to get to message 395.

Other arrangement schemes (such as by chronological order, by component or category, or by job responsibility) might be helpful to experienced users.

The following set of reference topics has an alphabetical organization:

Original

Reference information
- ABC log file
- Backup commands
- Database log file
- Extended recovery commands
- Fallback commands
- Runstat log file

The organization of the original reference topics is usable; however, the topics about commands are mixed with the topics about log files. Users who are looking for log files will need to look closely to find all of them.

Revision

Reference information
Commands
Backup commands
Extended recovery commands
Fallback commands

Log files
ABC log file
Database log file
Runstat log file

The revised reference topics are organized by category. The command topics are first, followed by the log file topics.

Try to use categories that the user is familiar with or can distinguish easily as shown in the revision. When information is divided into arbitrary or ambiguous categories, users must work harder to find what they need. Users can miss information or become frustrated with the organization.

The following topics are categorized arbitrarily into those that might be used by new users and those that might be used by advanced users:

Original

Beginning
Beginning query commands
Prompted query functions
Types of query reports
Sample tables
Launching the tutorial

Advanced
Advanced query commands
Query-by-example
SQL statements
Functions for customizing reports
Scalar functions
Data conversion functions
Customizing the tutorial

In the original structure, users must know which topics are "beginning" and which are "advanced" to know where to find the information that they need. The division between the two categories is subjective and depends on users' assessments of their own skill levels with the product. Some users might look for the topic "Prompted query functions" under the Advanced category, whereas other users could look for "SQL statements" under the Beginning category.

Revision

Functions
Functions for customizing reports
Functions to manipulate data
Scalar functions
Data conversion functions

InfoQuery commands

SQL statements

Querles
Prompted query
Query-by-example
Types of query reports

Sample tables

Tutorial
Launching the tutorial
Customizing the tutorial

In the revised structure, the two subjective categories are replaced by categories that make sense to all users.

In addition to organizing topics for quick retrieval at the outline level, you can help users by arranging the content of your topics to facilitate quick retrieval. If important information is difficult to find within a topic, users must spend extra time searching. In particular, reference information can often be presented in lists or tables for quick retrieval. See the retrievability guideline "Provide helpful entry points" on page 257 and the visual effectiveness guideline "Balance the number and placement of visual elements" on page 304 for details about how to make information easy to find within topics.

Other methods of aiding in quick retrieval are effective links between topics and effective index entries. These methods are discussed in Chapter 9 , "Retrievability," on page 245.

Separate contextual information from other types of information

Contextual information explains the product windows and the fields and controls that the user is working with. This information might include the syntax for a field, conditions for when a control is available, what a control does, or even how to use a product window. Contextual topics can be short and are meant to be read when users are viewing the product windows or controls.

The following task topic provides contextual information inside the steps of the task:

Original

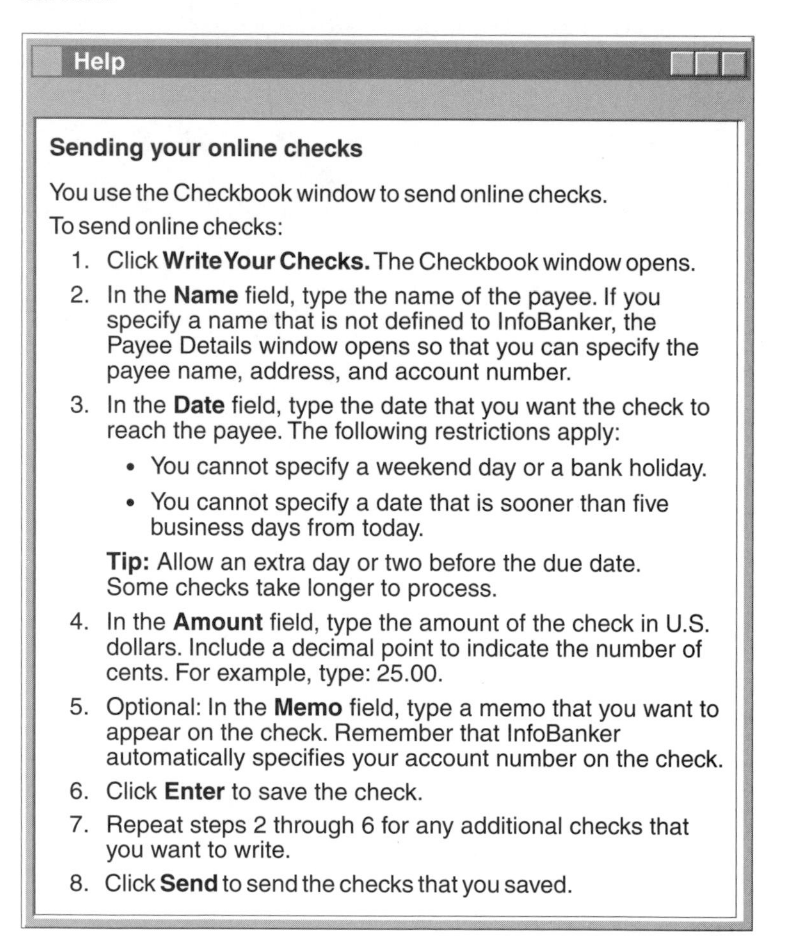

Help

Sending your online checks

You use the Checkbook window to send online checks.

To send online checks:

1. Click **Write Your Checks.** The Checkbook window opens.
2. In the **Name** field, type the name of the payee. If you specify a name that is not defined to InfoBanker, the Payee Details window opens so that you can specify the payee name, address, and account number.
3. In the **Date** field, type the date that you want the check to reach the payee. The following restrictions apply:
   - You cannot specify a weekend day or a bank holiday.
   - You cannot specify a date that is sooner than five business days from today.

   Tip: Allow an extra day or two before the due date. Some checks take longer to process.
4. In the **Amount** field, type the amount of the check in U.S. dollars. Include a decimal point to indicate the number of cents. For example, type: 25.00.
5. Optional: In the **Memo** field, type a memo that you want to appear on the check. Remember that InfoBanker automatically specifies your account number on the check.
6. Click **Enter** to save the check.
7. Repeat steps 2 through 6 for any additional checks that you want to write.
8. Click **Send** to send the checks that you saved.

The original task topic shows the contextual information for each control mixed in with the task steps. Users who want information about what to specify in the **Date** field need to read through the steps to find the information that is associated with that field.

Another problem with mixing contextual information and task steps is that some controls on a window are not actually involved in the task. Some windows might be used for multiple tasks and therefore contain controls that are used in one task but not all tasks for that window. In addition, some controls might be extraneous to any task; for example, the **Cancel** push button is rarely part of a task. If you rely on the task steps as the only method of explaining the controls on a window, you can end up either with some controls that are not explained, or with unwieldy tasks that have extraneous steps for unnecessary controls.

First revision

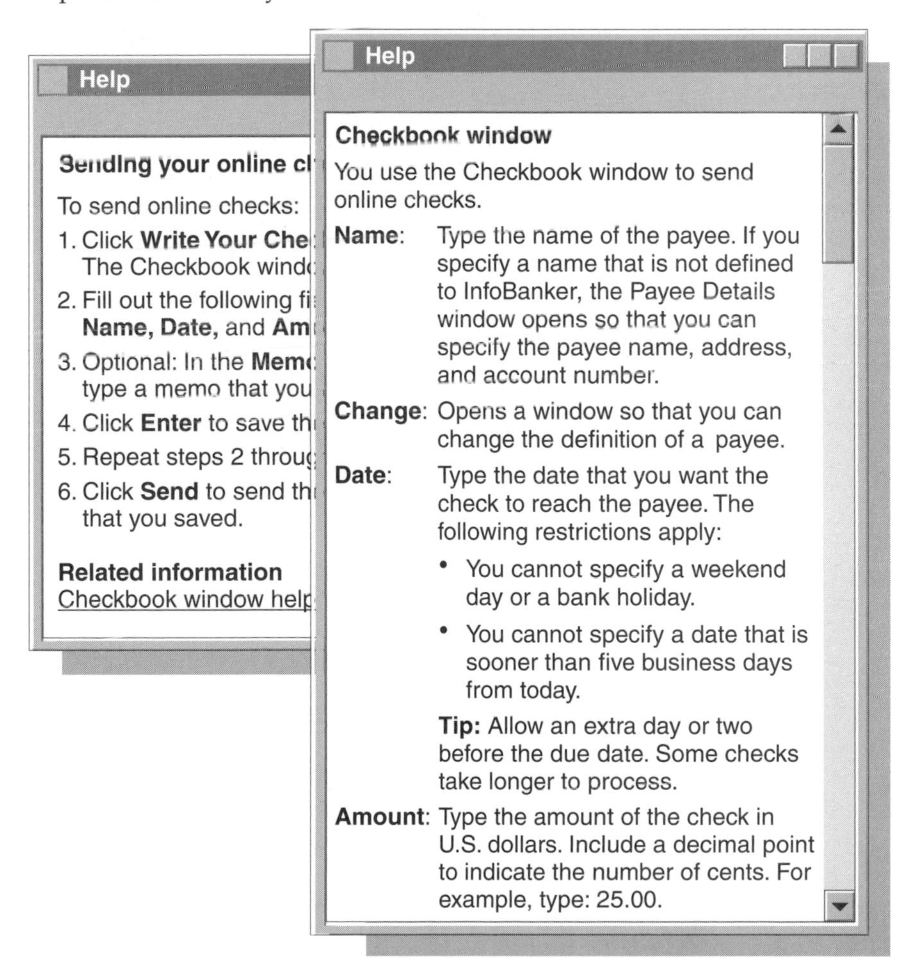

The first revision shows the contextual information separated from the task topic. The first revision includes contextual help for the **Change** push button, which was not mentioned in the original steps for sending online checks because changing a payee is a different task. The task topic in the first revision is now streamlined because it focuses more on the steps to accomplish the task and less on details about each of the controls that are used in the task. The user can use the task topic with the contextual help topic to fully understand both the task and the controls on the window.

In some cases, you can provide the contextual help by using pop-up help for each control. The following revision shows what pop-up help might look like for some of the controls on the Checkbook window:

Second revision

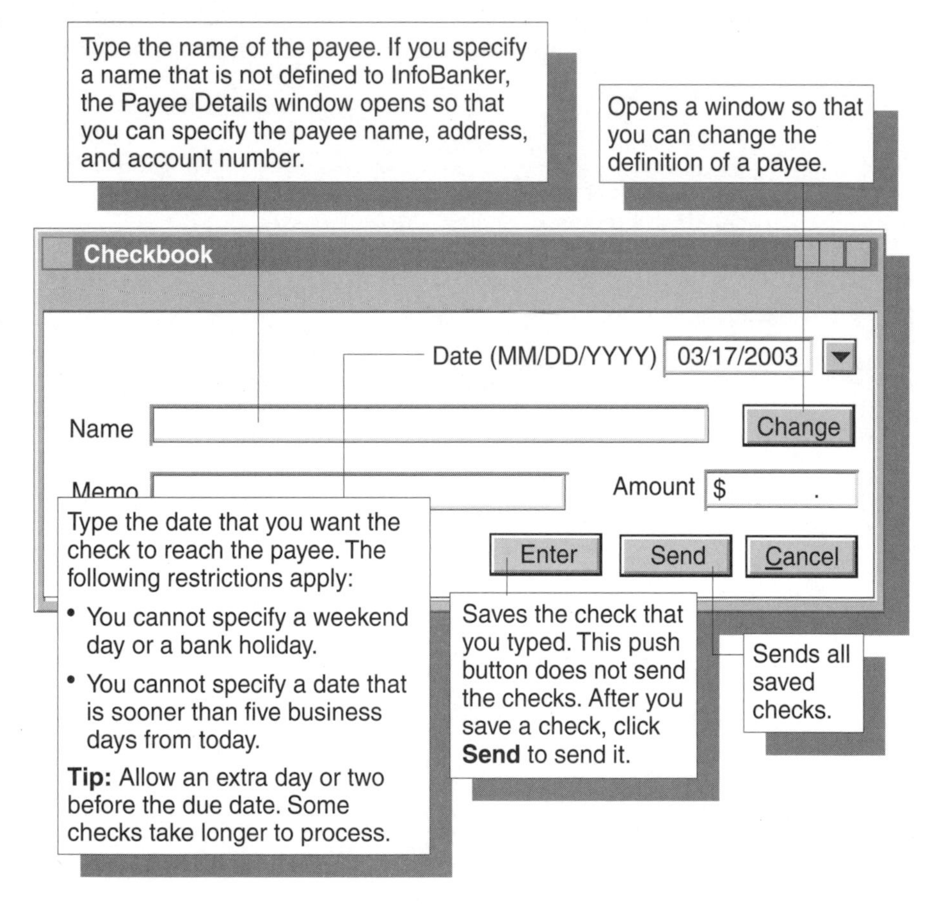

By providing contextual help for fields and controls in the user interface as shown in the second revision, the writer streamlines the task help as shown in the first revision, avoids the need to provide a separate contextual help topic, and presents timely information about the window's fields and controls to users who need it.

Determine what information users need for each product window and control, and provide that information in a format that is easy to find. Ensure that contextual information is available from the interface or in a topic that users can link to from the associated task topics.

Organize information consistently

Consistent organization helps users become familiar with the structure of information so that they can find what they need with confidence. When you present information consistently, users learn to predict what information they will find and where they will find it.

Organizing information consistently also helps ensure that the information is complete. Therefore, this guideline is similar to the completeness guideline "Use patterns of information to ensure proper coverage" on page 90.

The organization of standard content, such as getting-started information or a contextual help topic, should follow a similar pattern across related products. For example, if you include restrictions before the steps in some task topics and after the steps in other task topics, users might not notice some of the restrictions.

The following navigation pane shows topics that are organized inconsistently:

Original

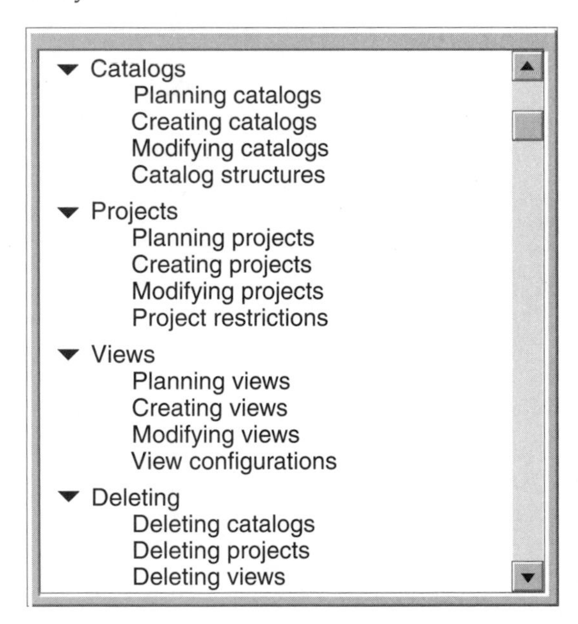

Because the information in the original navigation pane is organized by object type, the branch for deleting objects is out of place. Users who navigate with this navigation scheme learn to look for topics according to the type of object that they describe. These users don't expect to find the topics about deleting objects in a separate branch. For consistency, the topics about deleting objects should be presented with their appropriate objects as shown in the revised navigation pane.

Revision

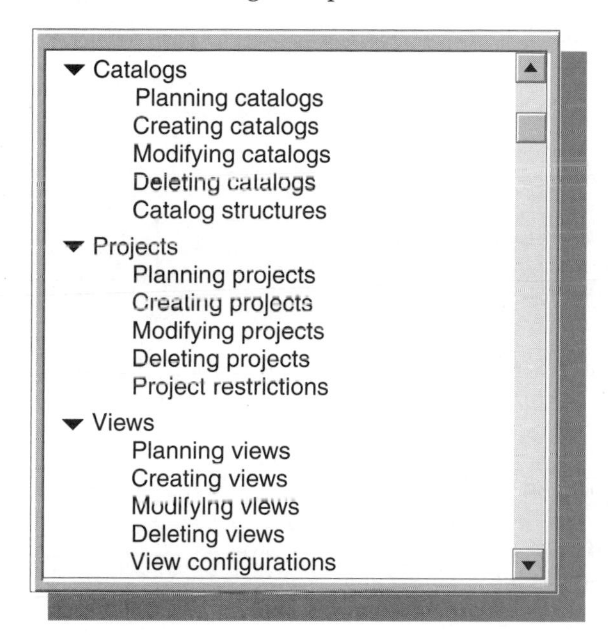

In the revised navigation pane, the topics are organized consistently by object. Your information center might not have a navigation scheme that is organized by object, but regardless of the navigation scheme that you provide, organize the topics within the navigation pane consistently.

Provide an appropriate number of subentries for each branch

If a navigation scheme contains a branch that has a single subentry, the navigation scheme might have an organization problem. If there is only one main thing to say at a branch level, a subentry is probably not needed.

The following book outline shows a chapter with one subsection:

Original

Chapter 8. File structure **201**
File structure table... 201
Chapter 9. Troubleshooting.................................. **205**

Chapter 8 in the original outline contains a brief introduction followed by one section. Therefore, the text in the "File structure table" section makes up most of the chapter.

Revision

Chapter 8. File structure table............................... **201**
Chapter 9. Troubleshooting.................................. **205**

Single subentries can sometimes replace the branch itself, as shown in the revised outline, where the subsection is now the chapter. The brief introduction in the original outline is either omitted or is added to the "File structure table" topic.

This outline shows a book structure, but the example applies to the organization of topics as well as to chapters and sections. Suppose that the original outline showed an information center navigation pane instead of a book outline. You would put the "File structure table" topic at the branch level instead of creating a branch with a single subentry for "File structure table." By removing unneeded introductions and branches, you can streamline your navigation structure.

Branches can contain single subentries when the subentry presents a topic that is separate from the branch yet logically subordinate to the branch. For example, if your tasks have long examples, you might decide to separate the examples into their own topics. In the navigation pane, you show the examples subordinate to the tasks hierarchically. However, if task A has only one example and no other subentries, its branch will have a single yet logically subordinate subentry.

A more common problem associated with single subentries is not including enough topics at the appropriate level, as shown in the following outline:

Original

Data security
  Access
Data integrity

The original outline shows only one subentry for data security.

Revision

Data security
  Access to the InfoDBase system
  Access to the InfoDBase database
  Access to the data source system
  Access to the data source database
Data integrity

In the revision, the writer divided the topic "Access" into four separate topics. "Data security" now has four subentries so users can choose the specific type of access to read about.

Ensure that all single subentries are logically subordinate to their branch. If they are not logically subordinate, remove extraneous introductions and branches. If the single subentries are logically subordinate, check to see whether more subentries are needed at that level. In addition to dividing an existing topic, other solutions include finding additional topics for a branch with a single subentry or removing the branch and adding the single topic to another branch.

Emphasize main points; subordinate secondary points

You can use emphasis and subordination to distinguish main points from supporting information. Users need a way to distinguish the highways from the back roads; they get lost when they can't keep track of the main points.

Emphasizing a main point involves one or more of the following techniques:

Placement

Placing topics or words first is the best way to emphasize a main point: in the heading or title, in the introduction, in the first sentence of a paragraph, in the first words of a sentence.

Introduction

An introduction, or overview, should announce to the user why subsequent information is important or how that information is relevant. Such an orientation places the subsequent information in context and thus makes it easier for users to learn or to understand the order of topics to come.

Repetition and reinforcement

Repetition keeps the user's mind on the point throughout the presentation. Effective reinforcement includes making a point in more than one way, such as by text and illustration, or by general statement and example.

Details

Details help emphasize what is important. Give users more details on important topics, and fewer details on topics that are less important.

The following paragraph seems to be about paragraphs at first, but it is actually about how to use markup tags:

Original

> Paragraphs can have their first lines indented, or not; they can be set to the left margin or indented under headings; they can be separated by one blank line, two blank lines, or none. But, in the source, they all begin with a tag designating a paragraph. Similarly, a highlighted phrase can be underlined when it is printed on one device, appear in italics when printed on another, and display in red in a window. It is still the same kind of element, a highlighted phrase, identified by the same highlighting tag. Thus the generalized markup tag identifies the kind of element that a portion of a document is.

Users need to read the whole original paragraph to determine what the paragraph is about, and some users might not grasp the intended topic at all.

First revision

> The generalized markup tag identifies the kind of element that a portion of a document is, not the way that the element is to be formatted. Paragraphs can be set to the left margin or indented under headings; they can be separated by one blank line, two blank lines, or none. But, in the source, they all begin with a paragraph tag. Similarly, a highlighted phrase can be underlined when it is printed on one device, appear in italics when printed on another, and display in red in a window. It is still the same kind of element, a highlighted phrase, identified by a highlighting tag. Therefore, when marking a portion of a document, ask yourself, "What kind of element is this?" not "How do I want this to look?"

The first revised paragraph begins with the topic sentence. It uses repetition and contrast to emphasize the main point.

Second revision

> The generalized markup tag identifies the kind of element that a portion of a document is, not the way that the element is to be formatted. For example:
>
> - Paragraphs are identified with a paragraph tag. They can be set to the left margin or indented under headings; they can be separated by one blank line, two blank lines, or none. But, in the source, they all begin with a paragraph tag.
> - A highlighted phrase is identified by a highlighting tag. It can be underlined when it is printed on one device, appear in italics when printed on another, and display in red in a window. It is still the same kind of element, a highlighted phrase.
>
> Therefore, when marking a portion of a document, ask yourself, "What *kind* of element is this?" not "How do I want this to look?"

The second revision uses a list to set off the examples more clearly and make the information easier to scan. It also uses placement and reinforcement in the list items to keep the focus on the main point.

When you give unwarranted emphasis to secondary information, you make that information seem more important than it is. Though a minor point might belong in the text, elaboration of the point probably does not.

In the following passage, a note label is used to call attention to information that is no more pertinent than the other information in the passage:

Original

> You can cancel registrations if you no longer need to copy data or if the registration is erroneous. You can cancel only the registrations that you created.
>
> **Note:** Canceling registrations ensures that the InfoDBase control tables are properly updated and prevents database errors.

The original passage gives unnecessary emphasis to the statement about the benefits of canceling a registration. That statement is no more important than the other two statements, yet the note label gives users the impression that it must be more important.

Revision

> You can cancel registrations if you no longer need to copy data or if the registration is erroneous. You can cancel only the registrations that you created. Canceling registrations ensures that the InfoDBase control tables are properly updated and prevents database errors.

In the revision, the information fits into the original paragraph without rewriting.

Note labels are a tempting method to use to insert new information into previously written topics. However, you'll soon find that your topics have a lot of notes and no flow and that users can't follow your main points. In addition, too many notes will dilute the effectiveness of the note label to visually emphasize a point. Take the time to fit the new information into the existing structure, or reorganize the information to accommodate the new information.

For information about visual emphasis, see the visual effectiveness guideline "Use visual elements for emphasis" on page 288.

Reveal how the pieces fit together

When users first encounter a piece of information, they usually want to see it as part of a whole. Like travelers, they need a map both at the beginning of a journey and as their trip progresses to keep their bearings in relation to the destination.

For a book (whether online or printed), users rely on the following elements as a map: table of contents, headings, introductions or overviews, and transitions. In information centers, online help systems, and Web sites, users can keep their bearings with the following aids:

- A navigation pane or site map
- Overview topics that convey how subtopics are related
- Transitions and links between topics that convey the relationship of one topic to another
- A *breadcrumb* trail that lists the topics that a user has visited
- Buttons that take users to the previous topic or to the next topic

The navigation pane or site map provides a map to those users who need a sense of the breadth of the set of information. However, because information centers, online help systems, and Web sites are not read linearly, you should provide overview topics that list related topics and show how those topics relate to each other. Users can return to these overview topics as necessary to link to other topics. In this way, users get a sense of the structure of the information and can read some of it linearly if they want to. Users also rely on effective transitions and links to convey the relationships of topics to one another.

All of these aids help users to verify their sense of the whole. If they can't verify it or if a piece doesn't fit, something is wrong or missing.

The following set of topics does not provide overview information for the tasks that are involved in using the sample program:

Original

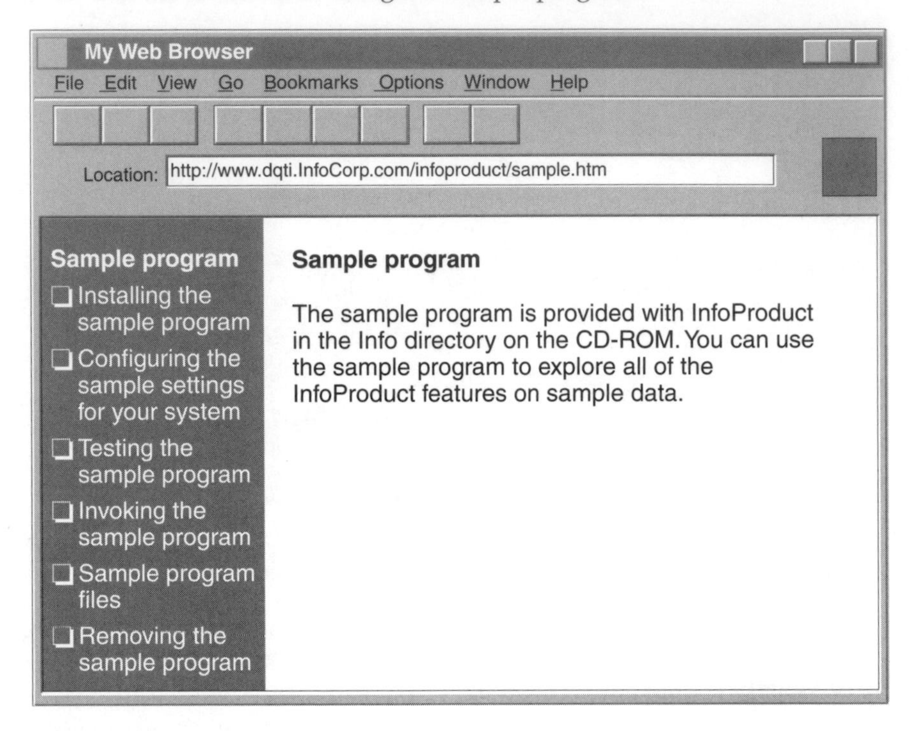

The original set of topics includes a concept topic about the sample program followed by the task topics in the order that the user should do the tasks, a reference topic that lists the sample program files, and a task topic about removing the sample program. The user cannot tell from the navigation pane whether the order is important or how the tasks relate to each other. The overview concept topic in the right frame does not explain the relationships of the subtopics.

Revision

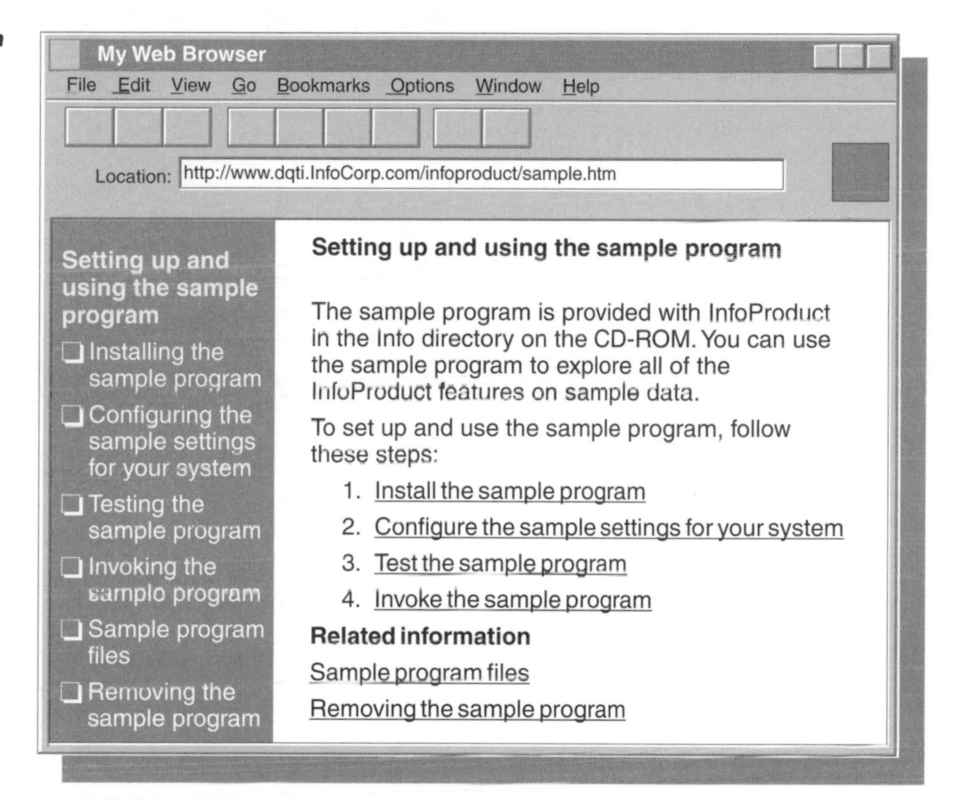

The revised set of topics includes an overview task topic that explains that the first four subtopics are all subtasks of the overview task and indicates the order in which to do them. The overview task topic provides information about the subtopic relationships that users cannot grasp from just the navigation pane alone.

Because topics are not always read linearly, you cannot assume that users have read topic A before reading topic B or that users who are reading topic A will know to go to topic B next. Therefore, introductions to tasks such as "Now that you have configured the receiver, you must configure the amplifier" are not sufficient. In the "Configuring the amplifier" topic, you need to mention that "Configuring the receiver" is a prerequisite task, and you need to provide a link to the "Configuring the receiver" topic. Do not assume that the user is reading linearly and has already read about configuring the receiver.

The same advice applies to other transitions between topics. Sentences such as "The next section explains how to configure the printer driver" assume that the user sees the sections in a certain order. A better way of mentioning a follow-on task might be to provide the following sentence in your topic: "After you complete this task, you must configure the printer driver."

The following passage shows the end of a topic that explains how to install a product:

Original

> 9. Press Enter. The InfoProduct installation process begins.

The original topic ends as soon as the installation process is started and doesn't point the users to the next place to go in the set of information. The users must figure out what to do next, if anything.

Revision

> 9. Press Enter. The InfoProduct installation process begins.
>
> You can now configure the InfoProduct settings for your system.
>
> **Related information**
>
> Configuring InfoProduct settings

The revised topic explains to users that they can now configure their system. The link to the related task is provided at the end of the topic for users who want to go to the next task.

Because users do not read information centers, online help systems, and Web sites linearly, users need to understand where they are in the set of information and where they can go. Navigation panes provide a hierarchical view of the structure of the set of information. Overview topics and transitions are valuable supplements to navigation panes because they help users understand where they are and what to do next. In addition, users rely on breadcrumb trails to help them return to overview pages so that they can take other links and explore without getting lost.

See the retrievability guideline "Facilitate navigation and search" on page 247 for more aids that you can use to help users keep their bearings.

In sum

Organization deals with much more than how you arrange your topics. Well-organized information never leaves the user feeling lost. The order of tasks is clear, transitions suggest follow-on tasks and lead the user in the right direction, topics are easy to find, topic content is arranged for quick retrieval, contextual information is not hidden, and the overall structure is always available to the user.

When you review technical information for organization, you can use the checklist on page 244 in two ways:

- ❑ As a reminder of what to look for, to ensure a thorough review
- ❑ As an evaluation tool, to determine the quality of the information

Based on the number and severity of items that you find, decide how the information rates on each guideline for this quality characteristic. You can then add your findings to "Quality checklist" on page 387, which covers all the quality characteristics.

Although the guidelines are intended to cover all areas for this quality characteristic, you might find additional items to add to the list for a guideline.

Guidelines for organization	Items to look for	Quality rating
Organize information into discrete topics by type.	• Topics are discrete. • Topics contain only one type of information: concept, task, or reference.	1 2 3 4 5
Organize tasks by order of use.	• Task topics are in order of use. • Task steps are in order of use. • Sufficient conceptual information introduces a task.	1 2 3 4 5
Organize topics for quick retrieval.	• Topics are easy to find. • Information is easy to find within a topic. • Order of topics makes sense to the user.	1 2 3 4 5
Separate contextual information from other types of information.	• Contextual information is easy to find. • Contextual information is available from the interface or linked to from task topics.	1 2 3 4 5
Organize information consistently.	• Sequence of topics is consistent. • Organization of standard content is consistent.	1 2 3 4 5
Provide an appropriate number of subentries for each branch.	• Subentries are logically subordinate to their branches. • Each branch has a sufficient number of subentries.	1 2 3 4 5
Emphasize main points, subordinate secondary points.	• Paragraphs emphasize main points. • Details support main points, but not subordinate points. • Note labels are rare and appropriate.	1 2 3 4 5
Reveal how the pieces fit together.	• Navigation pane, site map, or table of contents shows the organization. • Transitions and overviews provide topic relationships as needed. • Hierarchy of topics is apparent.	1 2 3 4 5

Note: The scale for the quality rating goes from very satisfied (1) to very dissatisfied (5).

Chapter 9

Retrievability

Knowledge is of two kinds. We know a subject ourselves, or we know where we can find information upon it. —Samuel Johnson

Retrievability is the characteristic of information that enables users to find specific information. Good retrievability ensures that users don't get lost in a maze of information. Users want to find what they need quickly and easily. Many elements contribute to good retrievability: obvious ones like tables of contents, indexes, links, and headings, and not so obvious ones like highlighting, revision markers, icons, and navigation panes. These elements are *entry points*—signposts that orient and direct users so that they can find the information that they want.

Imagine looking for a specific piece of information on a Web site, in a set of online information, or a book. Can you find that information if you look for it by using a method such as the search function, index, or table of contents? Does a search using a common synonym find the information? Do variations of the wording of the subject show up in the index? You need to make sure that the answer to each of these questions is yes.

If users can't find what they are looking for, it doesn't matter how accurate, complete, or task-oriented the information is. Making sure that your tasks are well documented without making those tasks easy to find is like sending someone to the store to get ice without telling him how to get there. Give that person directions to the store, and he can find what he needs. No matter what the final delivery medium is for the information that you are developing, you need to ensure that you provide your users with the right "directions" to find the information that they need.

To develop information with good retrievability, follow these guidelines:

- ❑ **Facilitate navigation and search.**
- ❑ **Provide a complete and consistent index.**
- ❑ **Use an appropriate level of detail in the table of contents.**
- ❑ **Provide helpful entry points.**
- ❑ **Link appropriately.**
- ❑ **Design helpful links.**
- ❑ **Make linked-to information easy to find in the target topic.**

Facilitate navigation and search

As more information moves to the Web and more help systems are browser based, you need to pay particular attention to navigation and online searching. By providing good navigation, you enable users to move between topics and within pages and Web sites. Good navigation also keeps users oriented and aware of their current location. Search mechanisms are what most users rely on above all other methods to find information online. When you develop technical information, be aware of the mechanisms that are available in the system in which your information is delivered.

On all pages, provide navigational links, such as a link to a higher-level page, main or home page, or table of contents. For example, many Web sites and online help systems use a navigation pane, traditionally on the left, which remains no matter where the user goes within the same set of information. The navigation pane can be connected to the content area on the right, so that it scrolls out of view as the user scrolls down, or it can be separately scrollable, so that it always remains in view.

If the navigation pane is not displayed in lower-level topics, provide navigational links at both the top and bottom of the page, such as a "Contents" link at the top and a "Top of page" link at the bottom. These links reduce the need for scrolling and help users maintain their frame of reference.

The following Web page doesn't tell the user what product the page supports and doesn't provide a map of the Web site for additional information:

Original

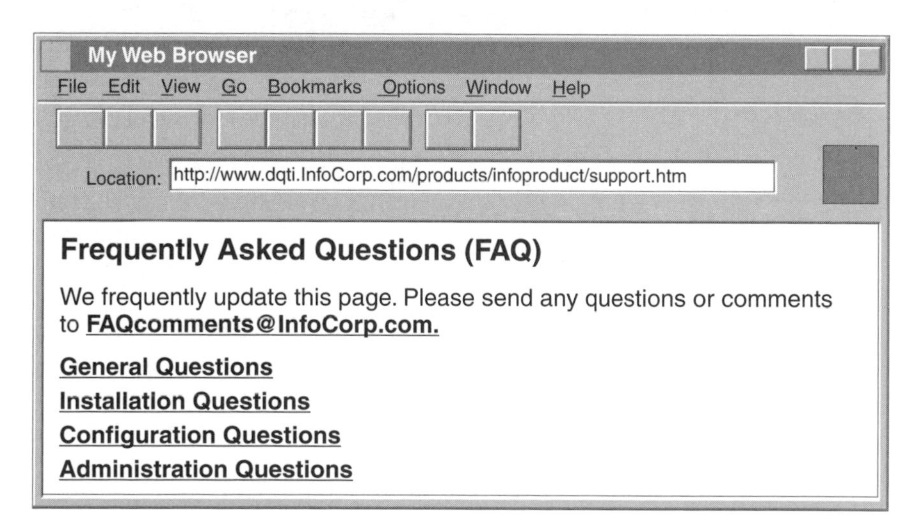

Users who find this Web page can't tell where they are just by looking at the content of the page. In addition, they can't link to the product's home page to find out more about the product or the company.

Revision

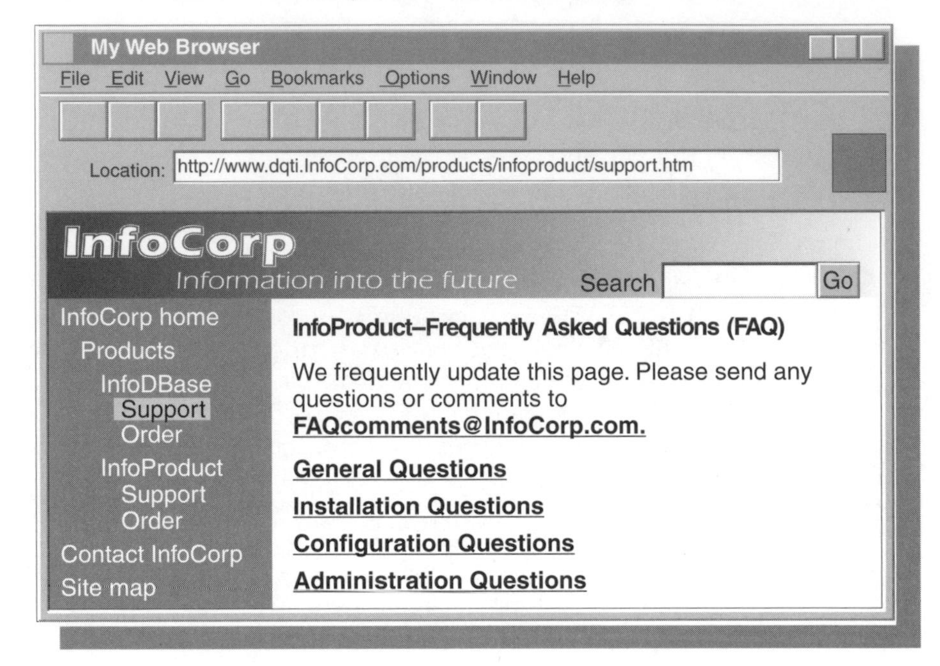

The revised Web page provides the company's name in the banner and the product name in the heading so that users know which product FAQ page they are reading. This page also provides navigational aids so that users can explore the other pages in the Web site and see where they are in relationship to the other pages. The search field and site map enable users to search the Web site for other information.

When you write help or Web information, consider the type of search tool that will be used. Some search tools might search only headings and defined keywords, while others might do a full text search. Make sure that the search tool can find the key topics, tasks, and terms in your information.

For Web search engines, you can use meta tags, which allow you to enter keywords and other information that Web search engines will find. Make sure that the information that you include in such meta tags is what you expect users to look for.

Most online indexes include search functions. Make sure that your index is designed to work well with the search function, as described in the following guideline.

Provide a complete and consistent index

In technical information, whether online or printed, in help systems or on the Web, users often go to the index first to look for information. You need to predict what information users will look for, and what words they might expect as pointers to that information.

As explained in the completeness chapter, completeness in technical information means neither too much nor too little information, but instead, just the amount that users need to complete their tasks. Indexes in technical information are occasionally too long, and online indexes sometimes have too many variations on an index entry. But it's far more common for indexes to be too short, often because indexing is left until the end of the project, when deadlines are pressing. An index needs to have entries for every important subject that is covered in the information.

Verbs can be effective index entries, especially when they are specific and describe a user task, such as "copying," "installing," or "searching." However, a common mistake in indexing is to use generic verbs followed by nouns as secondary entries, as in the following set of entries:

Original

```
using
  CGI  105-115
  data controls  142
  forms  93
  HTML  9
  Perl  105
```

Revision

CGI
  example 114
  when to use 105
  ⋮
data controls
  description 142
  example 144
  ⋮
forms
  description 93
  check boxes 96
  submit buttons 98
  ⋮
HTML
  overview 9
  validating 35
  ⋮
Perl 105

Users are unlikely to look for *using* in the index, no matter what information they are looking for, so the original entries are mostly a waste of time and space. Users are more likely to look for a specific noun or verb instead of a generic verb, so the revised entries are likely to help users find the information they want.

In general, use specific terms and avoid generic ones when you are creating an index. Some words can be specific or generic, depending on how they are used or modified. The word *examples* can be a good primary index entry if you have no more than a few examples in your document. But it's too generic if you have hundreds of examples, and in that case you're better off indexing examples under the specific topic that they apply to. Other words that can be specific or generic depending on the context include *creating*, *updating*, *commands*, and *messages*.

The words or phrases that users think of might not be the same ones that are used in the information, and not all users think alike, even about the same topic or task. Therefore, an index should contain synonyms for its entries. For example, to index a topic about font size, you would create an entry under *font*, and an entry under *typeface*.

Index entries that point to a specific term are usually more helpful than index entries that provide a *See* reference to another term. Although linking from a *See* reference to the appropriate term means one extra step for users, providing a *See* reference is much better than expecting users to scroll through the index or enter the preferred term to find the index entries that they expect.

In the following set of entries, users can click on the word *fonts* in the *typefaces* entry to go immediately to the entries for *fonts*:

fonts
- ❑ adding
- ❑ changing
- ❑ enhancements
- ❑ foreign languages
- ❑ installing

. . .

typefaces. *See* fonts

Printed indexes benefit from the use of variations of wording, so that users looking, for example, for how to copy a file, can find the topic "Copying files" by going to either *copying* or *files* in the index:

copying
- ❑ data from tables 121
- ❑ files 32
- ❑ tables 137

. . .

files
- ❑ adding to object tree 49
- ❑ copying 32
- ❑ creating 36

In online information, variations can also be useful, but in indexes with search functions, too many variations can overload the index and return too many results when users search for a given word. The following entries from an online index show too many variations:

Original

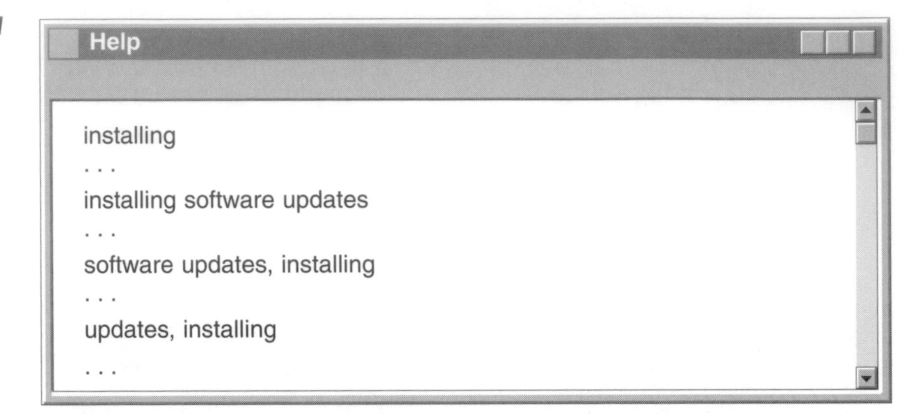

Revision

In the original excerpt, a search for *installing* returns four results, and a search for *software updates* returns two results, with all of these results pointing to the same place.

In the revised excerpt, users will find the entry whether they search for *software updates*, *installing*, or *updates*, so the single entry is all that is needed in an index with a search function. If there is no search function, you should create synonyms and variations as for a printed index.

A good index adheres to indexing guidelines that make your index entries consistent and easy to use. The following guidelines can help ensure that your index is complete and correct.

Guidelines for online and printed indexes

- ❑ Index every task, concept, and reference topic, and important subjects within topics. Include at least one index entry for every topic.
- ❑ Start every index entry with a specific noun or a specific verb. If necessary, add a modifier to the noun or verb to make it more specific. Whether you put the modifier before or after the noun or verb depends on how users would probably look up the entry.

- ❑ Avoid index entries that start with a generic term that users are unlikely to look up, such as "changing," "managing," "tasks," or "using."
- ❑ If you use secondary or tertiary entries, put at least two of them at each level. (This requirement might be hard to meet if you are optimizing entries from several information units to be included in a master index.)

Guidelines for online indexes only

- ❑ Do not index contextual help. Such subordinate help topics do not need index entries because they make sense only when the user is viewing the actual product window.
- ❑ If you include variations of index entries, do so sparingly, and consider the search engine that will be used.

Guidelines for printed indexes only

- ❑ Include approximately one column of index entries for every 10 pages of text.
- ❑ Include plenty of synonyms and variations (*permutations*) of your index entries.
- ❑ If you use subentries, put page numbers on only the lowest level of subentry.
- ❑ If an entry has page numbers, it should have no more than two.

Indexing is a complex subject—many people make a living as professional indexers. For references to more detailed information about indexing, see "Easy to find" on page 407.

Use an appropriate level of detail in the table of contents

Information is often subdivided to many levels, with headings for each subdivision. Such a structure is useful to show a hierarchy of topics, but you do not need to include every heading of every subdivision in the table of contents. This practice usually produces a long table of contents that makes general subjects harder to find, and hides rather than reveals the organization of the information. A table of contents should provide more than the highest level of headings, but allow users to easily see the top-level hierarchy.

For Web documents such as PDF files that display the contents in a separate, scrollable frame, using expandable sections gives you the flexibility of including the lowest level of headings. Users can see at a glance the top-level headings, and choose to see lower-level headings.

Expandable links are especially useful when the table of contents is long and complex. When you provide expandable links, choose the top-level entries carefully, so that users can predict what lies underneath each one and quickly grasp the overall organization. For example, top-level entries such as "Introduction," "Concepts," and "Restrictions" are not likely to be helpful to users.

For HTML documents that do not have a separate navigation pane, provide major headings in the table of contents, but do not show more than two or three levels of headings in the table of contents, unless you can use expandable sections.

The following reference information is a single-frame document on one Web page:

Original

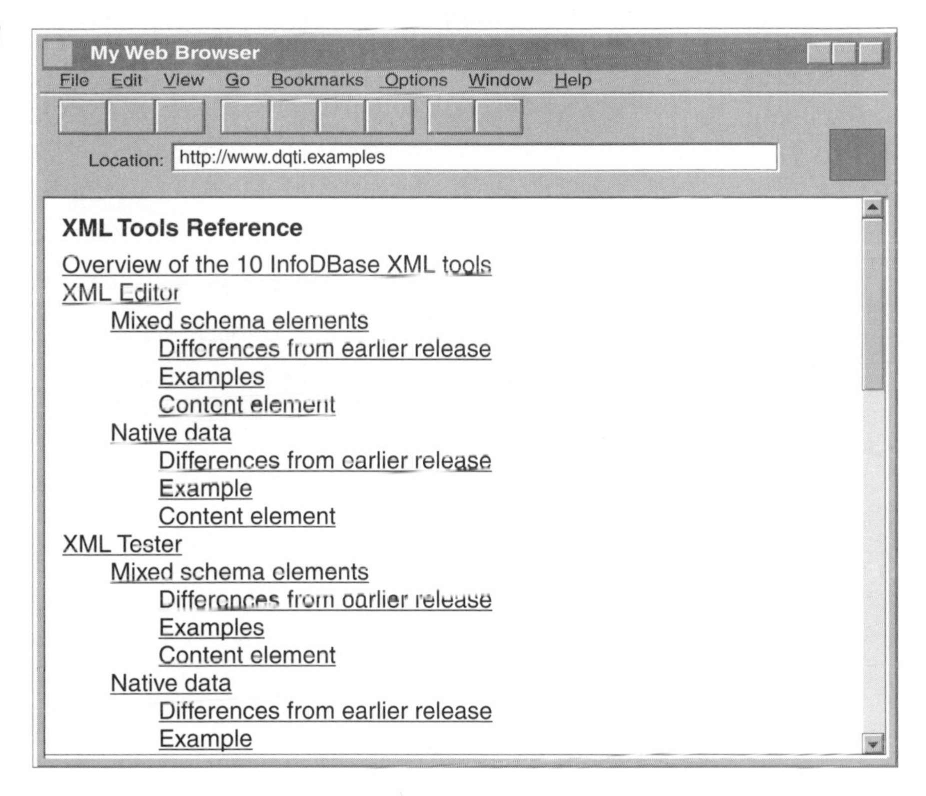

The original table of contents provides links to the lowest-level headings in the document. Such links are not necessary in reference information when each section contains similar information. Such detail also makes the contents too long—with this much detail, users must do quite a bit of scrolling to see the entire table of contents.

Revision

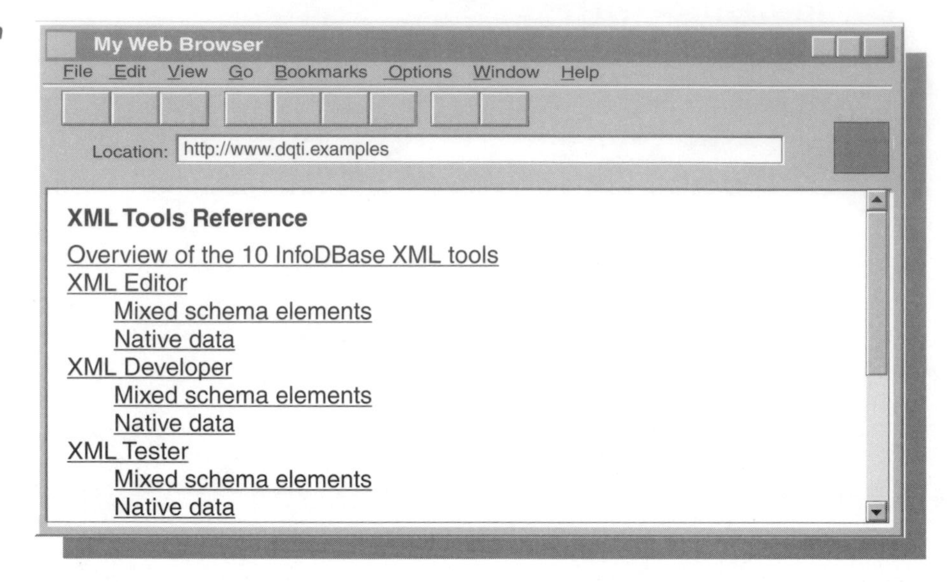

The revision removes the third-level headings, and results in a much shorter table of contents. Users can easily find the heading for a specific tool, as well as see the hierarchy of topics.

For online help, the table of contents should include every major help topic, but not contextual help topics. A user who looks for help on a specific product window is probably looking at that window at the time, and can get contextual help there.

For printed information, use approximately one table of contents page for every 50 pages of text. For long books (more than 400 pages), a ratio closer to one table of contents page for every 100 pages of text might be more appropriate.

A table of contents should have an easy-to-understand heading, such as "Contents" or "Page contents" (on the Web). Do not use an ambiguous heading such as "Quicklinks"; make sure that links and the headings that contain them are specific.

Provide helpful entry points

Few things in complex technical information are more discouraging than large blocks of text, unbroken by headings, graphics, or other visual elements. User expectations also change as technology changes. With the pervasiveness of the Web, users are accustomed to jumping from site to site and from topic to topic to find what they want. Users are less tolerant of large blocks of text and more adapted to searching for only the topics that interest them and to reading small chunks of information.

To support scanning, break up long blocks of text and provide entry points such as headings, labels, and lists. Aim for three to five sentences per paragraph, but also remember that the ease of understanding depends on the content and length of the sentences, as well as the complexity of the sentence structure.

The following topics show the difference between a solid block of text and the same text reorganized into a more retrievable format:

Original

Creating a dimension

Because this is the first dimension that you are creating for your plan, the InfoWrite editor is empty, ready for you to type your definitions. First, type a name for the dimension in the **Dimension** field. The name can be up to 12 characters, including uppercase or lowercase letters, but must contain no spaces. The first character must be alphabetic. Next, in the **Level** field, type the number of the highest level in the dimension hierarchy. The highest level is the level at the top of the dimension hierarchy. It has the lowest number in the dimension definition. It is good practice to leave gaps between the level numbers so that intermediate levels can be added later. For example, in a dimension that has four levels, use level 10 as the highest, then levels 20, 30, and 40 as the next levels. Type the name of the element that is at this level in the **Contents** field. The name must be unique and can have up to 24 characters, including uppercase and lowercase letters and spaces. Press Enter to add it to the list of dimension elements.

In the original text, several kinds of information are run together in one overwhelming paragraph. But the paragraph actually contains steps for a procedure (with a tip and example), rules for naming, and conceptual information.

Revision

Creating a dimension

Because this is the first dimension that you are creating for your plan, the InfoWrite editor is empty, ready for you to type your definitions.

To create a dimension:

1. In the **Dimension** field, type a name for the dimension.

   The name can be up to 12 characters, including uppercase or lowercase letters, but must contain no spaces. The first character must be alphabetic.

2. In the **Level** field, type the number of the highest level in the dimension hierarchy.

   The highest level is the level at the top of the dimension hierarchy. It has the lowest number in the dimension definition.

   Tip: Leave gaps between the level numbers. You can then add intermediate numbers later.

   Example: In a dimension that has four levels, use level 10 as the highest, then levels 20, 30, and 40 as the next levels.

3. In the **Contents** field, type the name of the element that is at this level.

   The name must be unique and can have up to 24 characters, including uppercase and lowercase letters and spaces.

4. Press Enter. The new element name is added to the list of dimension elements.

The revised text clearly shows the task steps by using an ordered list, so that users can more easily see the actual procedure. The rules and conceptual information are set off as separate paragraphs, and the tip and example now have labels to make them easier to spot.

Headings are the most common element used to break up information, but avoid adding unnecessary headings. Just because some headings are good doesn't mean that more are better. Too many headings can distract from each other and disrupt the organization and flow of the information, particularly when the heading levels change frequently.

Just as important as the quantity of headings in a document is their quality. Headings must be specific, giving a brief, clear description of the topic. Avoid abstract headings such as "Considerations" or "Important concepts."

Headings should contain the distinguishing information at the start and less important or repetitive information at the end. When the first words of the heading contain the important information, users can spot the information easily when they scan a table of contents. Also, in some contexts, such as bookmarks, application icons, and the Microsoft Windows task bar, often only a word or two is visible. For example, if a user has several browser windows open, the task bar will show just the start of each page title.

The following headings are not helpful:

Original

Restrictions
About XSL
Overview
Chapter 1. Background
File structure
Using InfoFinder

Revision

Restrictions for using InfoInstaller on UNIX
XSL style sheets
Topic maps - Overview
Chapter 1. Terminology process background
Structure of the DADX file
Finding information in the InfoCenter

The original headings all lack either enough distinguishing detail or a focus on the actual content. Each revised heading clearly indicates the information that users will find in the topic. For information about task-oriented headings, see the task orientation guideline "Use headings that reveal the tasks" on page 31.

Link appropriately

Links are essential in technical information. (This guideline uses the word *links* to refer to both online links and cross-references in printed information.) If you use an appropriate set of helpful links, you can increase the retrievability of the information. However, if the links are not helpful, or if you use too many or too few links, the user's reading is interrupted without good reason.

Links provide:

- Information about related topics
- Navigation within the current topic or document
- Navigation within a Web site or corporate Web domain
- Access to related information

Decisions about when to provide a link and when to repeat information affect completeness as well as retrievability. The guideline "Repeat information only when users will benefit from it" on page 96 deals with this issue from a completeness perspective.

You also need to make sure that links are correct, as described in the accuracy guideline "Check the accuracy of references to related information" on page 69.

Add links only when they are appropriate, and use them consistently. Because Web pages can take several seconds to load, unhelpful or redundant branches are especially tiresome to Web users.

In online information, you can provide navigation aids (such as previous and next buttons) for users without cluttering the topic with unnecessary links. You can also structure documents so that links to related topics appear consistently at the bottom or top of the window or on one side of a split window.

The following topic is cluttered with links inside the main text:

Original

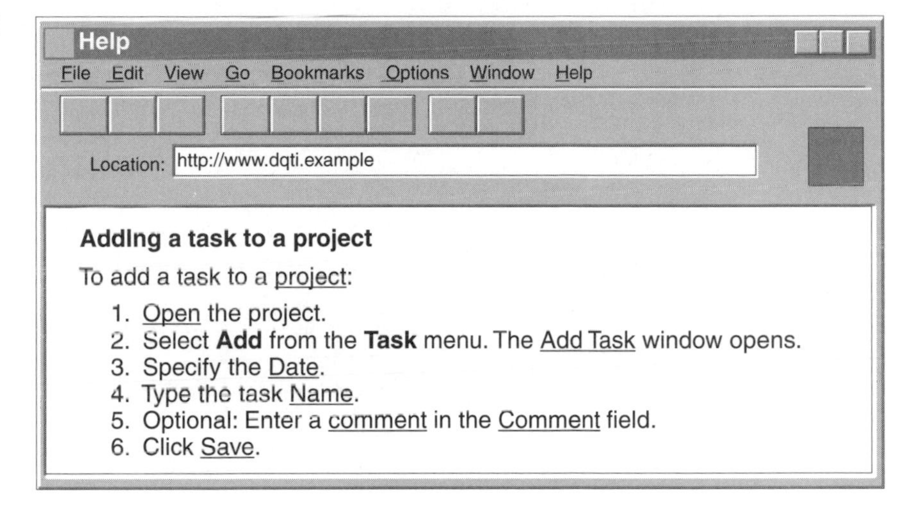

The original topic contains too many links to other information, even information that is redundant or unhelpful. The link for Open goes to a brief topic that explains how to open a project. The links for the Date, Name, and Comment fields each go to a topic that explains the format of input for those fields. Linking to these explanations might frustrate, rather than help, users.

First revision

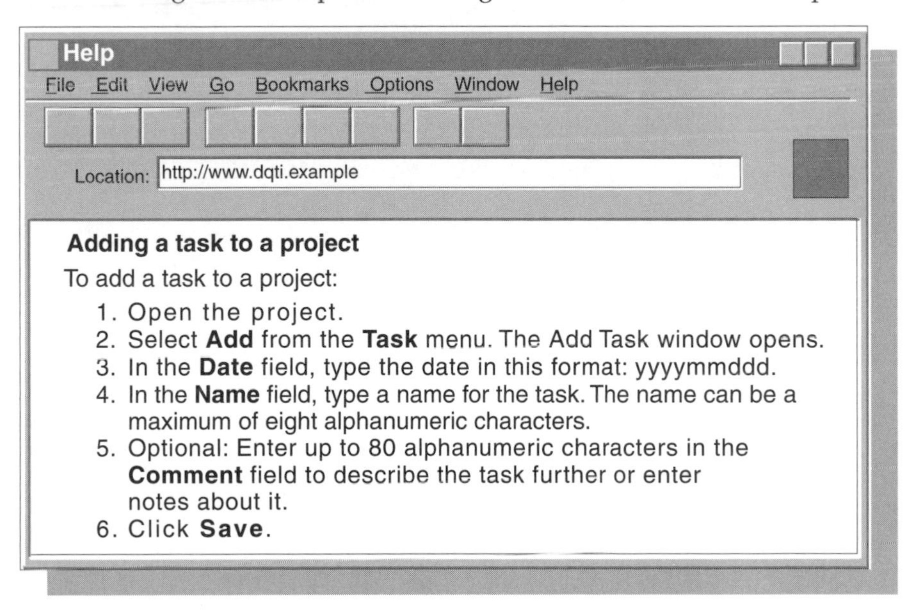

The first revised topic uses links only for relevant information. Some of the information that was linked to in the original topic is now included in the steps in the first revision.

Second revision

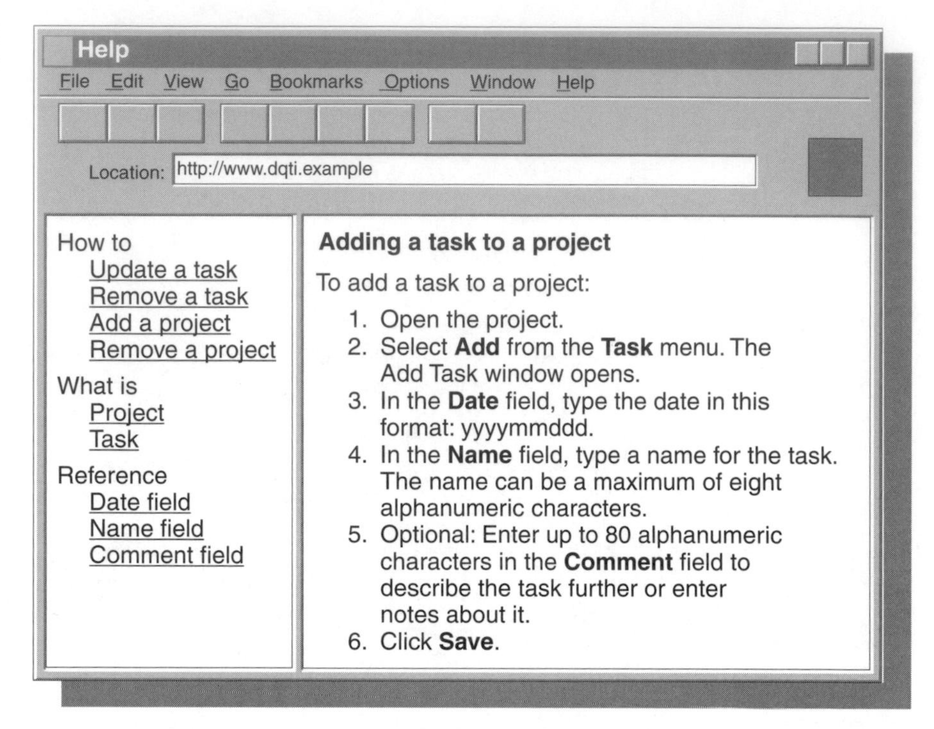

The second revised topic uses a left pane to provide links to related information. Users can take the links on the left if they are interested in getting more information about specific items—they do not need to decide whether to interrupt their reading (and their task) to link to other topics.

Users of printed information rarely interrupt their reading to refer to remote information, necessary or not. They resist even a branch that they are told to take, and read on until they find that they really can't proceed without taking that branch. Often, they waste time by trying to go ahead. For example, many users will ignore a reference to a figure that is only as far away as the facing page. But when the figure appears *on the same page*, right after the reference to it, almost all users will look at the figure immediately.

The link in the following passage is to a topic that explains the syntax for a command:

Original

Administrator	Can authorize other users and can register the data. The administrator is the owner of the user ID that runs the ABCONTROL command during installation. See "ABCONTROL command" on page 129 for more information about the command.
User	Can enter data into a subset of tables, as authorized by the administrator.

Revision

Administrator	Can authorize other users and can register the data. The administrator is the owner of the user ID that runs the ABCONTROL command during installation.
User	Can enter data into a subset of tables, as authorized by the administrator.

In the original passage, the user is reading about the difference between an administrator and a user. The user doesn't need to know the syntax of an installation command. The original passage provides the cross-reference in case a user might be immediately interested in that information, which is unlikely in this context. The revised passage saves most users the trouble of branching to unnecessary information. Any users who are especially interested in the ABCONTROL command can look it up through the index or search function.

Embedded links are links that appear in mid-sentence or mid-paragraph. Such links are disruptive because the user must decide whether to go immediately to the information, to come back to the link later, or to ignore the link entirely. Avoid using more than one embedded link for every two paragraphs of text. If you need more links, separate them from the text and put them in a separate section. If your list of related topics is long, consider the clarity guideline "Keep lists short" on page 129.

Design helpful links

Another consideration for providing appropriate links is the wording and style of the actual links. You need to be sure that the text of the link conveys the type of information that is being referred to. Before taking a link, users should understand, for example, whether the linked-to information defines a term or gives steps in a task.

The following guidelines can help you to design helpful links and cross-references:

- ❑ Ensure that a link describes the information that it links to.
- ❑ In similar links and cross-references, emphasize the text that is different.
- ❑ Minimize the effort that is needed to reach related information.
- ❑ Avoid redundant links.

Ensure that a link describes the information that it links to

Many Web pages contain single-word links such as "Resources," "Community," or "Offerings." Similarly, a link of "Related concept" does not tell the user what information to expect. In some cases, the context can make links like these meaningful to a specialized audience, but more often they cause confusion.

One type of unhelpful link often appears at the end of a paragraph or topic: "See <*title*> for more information." Such links give no indication of why the user should seek out the linked information. Users might have questions about what they are reading, but this cross-reference does not provide enough information for them to understand that they might find the answers in the linked information. A helpful cross-reference is something like: "For information about the XLINK language syntax, see ...." This sentence tells the user exactly what type of information the referenced topic contains.

Avoid creating links in text that say "Click here to read more about...." Links should keep the user's focus on the content, not on the links or the underlying structure of the topic. Instead of using clunky phrases such as "click here," make each link a meaningful noun phrase or verb phrase, such as:

- ❑ Visual programming tutorial
- ❑ FTP tools—software requirements
- ❑ Installing InfoProduct
- ❑ Download Version 3.0 beta code

In printed information, if the cross-reference is within the same book, make sure that it includes a page number. If the reference is in a different book, make sure that the text that you use to describe the related information helps the user to find that information quickly in the other book. Use the same wording that is used in the table of contents or index (or both) of the other book.

In similar links and cross-references, emphasize the text that is different

When you use lists of links with similar text, put only the words or phrases that are different inside the link, instead of the entire phrase. An even better solution is to move the redundant text to the top of the list.

The following links all start with the same words:

Original

- Software requirements for Linux
- Software requirements for Windows NT
- Software requirements for Windows 2000

Revision

Software requirements

- Linux
- Windows XT
- Windows 2000

In the original set of links, the words "Software requirements" are repetitive, and the operating systems are not immediately apparent. The revision moves the common text to a heading for the links, so that the important information is more prominent.

Minimize the effort that is needed to reach related information

For online information, make sure that users can find what they are looking for with as few links as possible. Most users would rather scroll through a list to find a specific topic than click through various levels or locations to find the right topic. When users need to jump to different pages or topics, they can become disoriented because their frame of reference changes. Scrolling, however, is two-dimensional and allows users to maintain their frame of reference.

Cross-references in printed information have various degrees of remoteness: the reference might be to the facing page, to the other side of the page, to a specific page in the book, to a topic in the same book with no page number specified, to another book closely related to the one in hand, to a book for another product, or to online information. Consider carefully whether the user needs this reference to another topic in context. One solution is to add cross-references at the end of a topic with a heading such as "For more information." Then you don't interrupt users' reading, but you still provide links to help them learn more.

Avoid redundant links

In a topic, two links that go to the same target topic can confuse or frustrate users. For example, a Web page might link to a technical topic by using its heading, while another link in a sidebar on the same page might link to the same topic using different text. This practice often wastes space and does not help users, as shown in the following topic:

Original

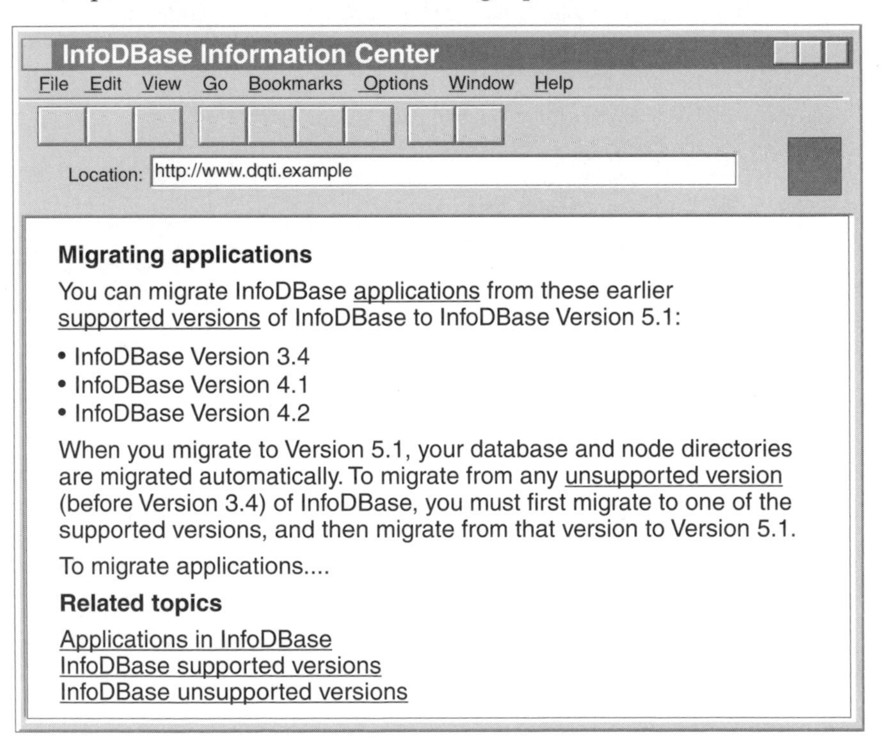

The topic contains links in the introductory paragraphs to information that is of interest to users who are migrating applications, but it links to the same information in the "Related topics" section at the end of the topic.

Revision

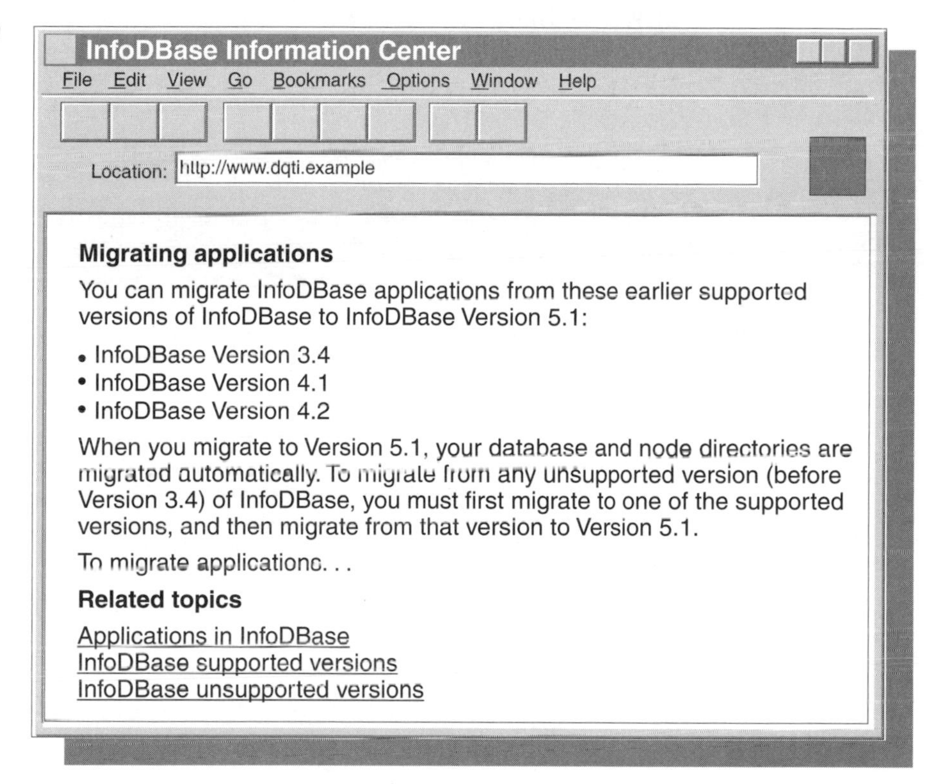

The revised topic removes the redundant links from the introductory material, leaving the links in the "Related topics" area.

Redundant links can be useful in some cases, however. A common and beneficial practice for Web sites is to provide links in a navigation pane, usually on the left, and the same links, with the same wording, at the bottom of the page. Users can find links easily and quickly.

Make linked-to information easy to find in the target topic

To make your information retrievable, you need to consider the different ways that users might get to a topic, and anticipate what pieces of information they will be looking for when they find that topic.

For example, a user might get to a help topic from the index or the table of contents, by using the search function, by linking from another help topic, or from a context-sensitive link from the product. In a printed book, a user might get to a page from the index, the table of contents, or a cross-reference. In both cases, you need to make the linked-to subject stand out in the target topic.

If your online help system links to the top of a topic instead of to a specific place within the topic, then put the linked-to information in the heading or first sentence of the topic. If this placement is not feasible, highlight the pertinent information in the topic, by using an icon, a subheading, or a label, for example.

For Web pages, consider the most likely sources of links into your page. A user might link from other pages within your Web site, from other pages controlled by your organization, from an application, or from an Internet search engine. In each case, try to anticipate what information users will be looking for, and then use highlighting, color, icons, labels, figures, or headings to make that information stand out.

The following help topic is displayed when the user selects "changing fonts" in the help index:

Original

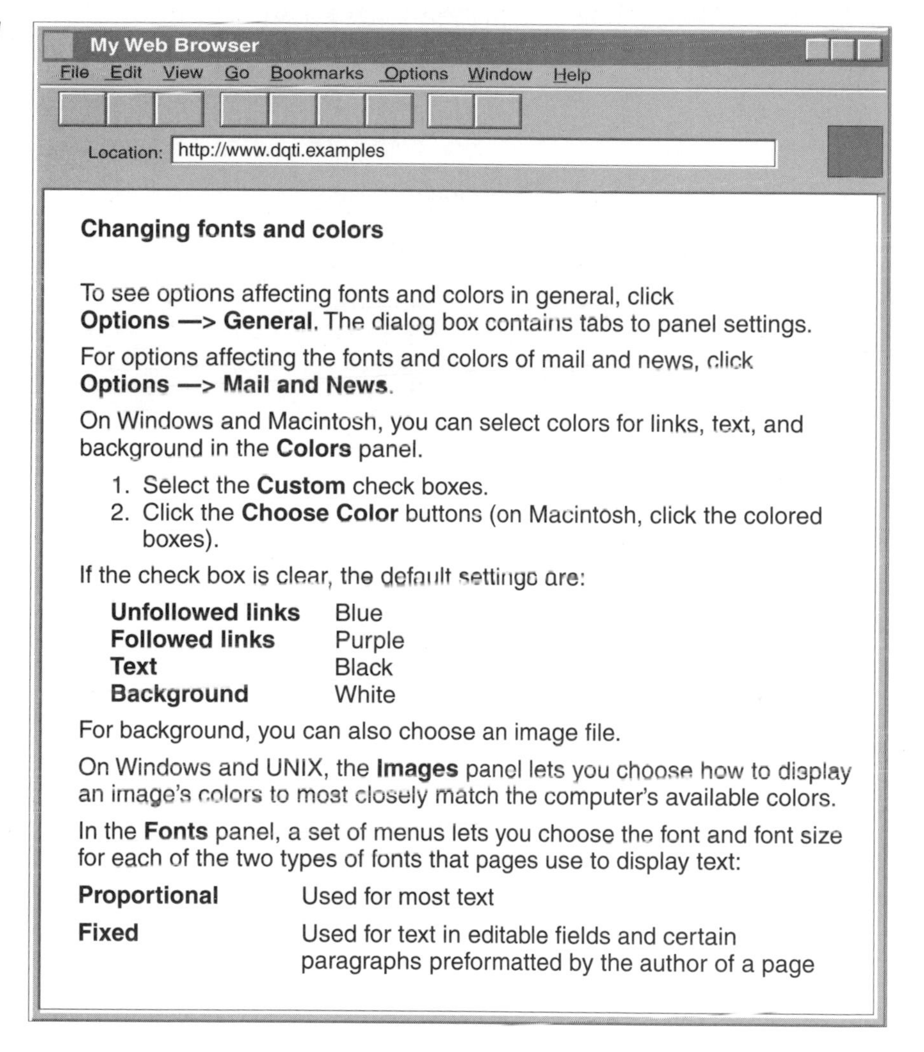

My Web Browser
File Edit View Go Bookmarks Options Window Help
Location: http://www.dqti.examples

Changing fonts and colors

To see options affecting fonts and colors in general, click **Options —> General**. The dialog box contains tabs to panel settings.

For options affecting the fonts and colors of mail and news, click **Options —> Mail and News**.

On Windows and Macintosh, you can select colors for links, text, and background in the **Colors** panel.

1. Select the **Custom** check boxes.
2. Click the **Choose Color** buttons (on Macintosh, click the colored boxes).

If the check box is clear, the default settings are:

Unfollowed links	Blue
Followed links	Purple
Text	Black
Background	White

For background, you can also choose an image file.

On Windows and UNIX, the **Images** panel lets you choose how to display an image's colors to most closely match the computer's available colors.

In the **Fonts** panel, a set of menus lets you choose the font and font size for each of the two types of fonts that pages use to display text:

Proportional	Used for most text
Fixed	Used for text in editable fields and certain paragraphs preformatted by the author of a page

In the original topic, the information about changing fonts is at the end. In the revised topic, headings guide users to the information that they're looking for. Also, the order of sections corresponds to the order in the heading.

Revision

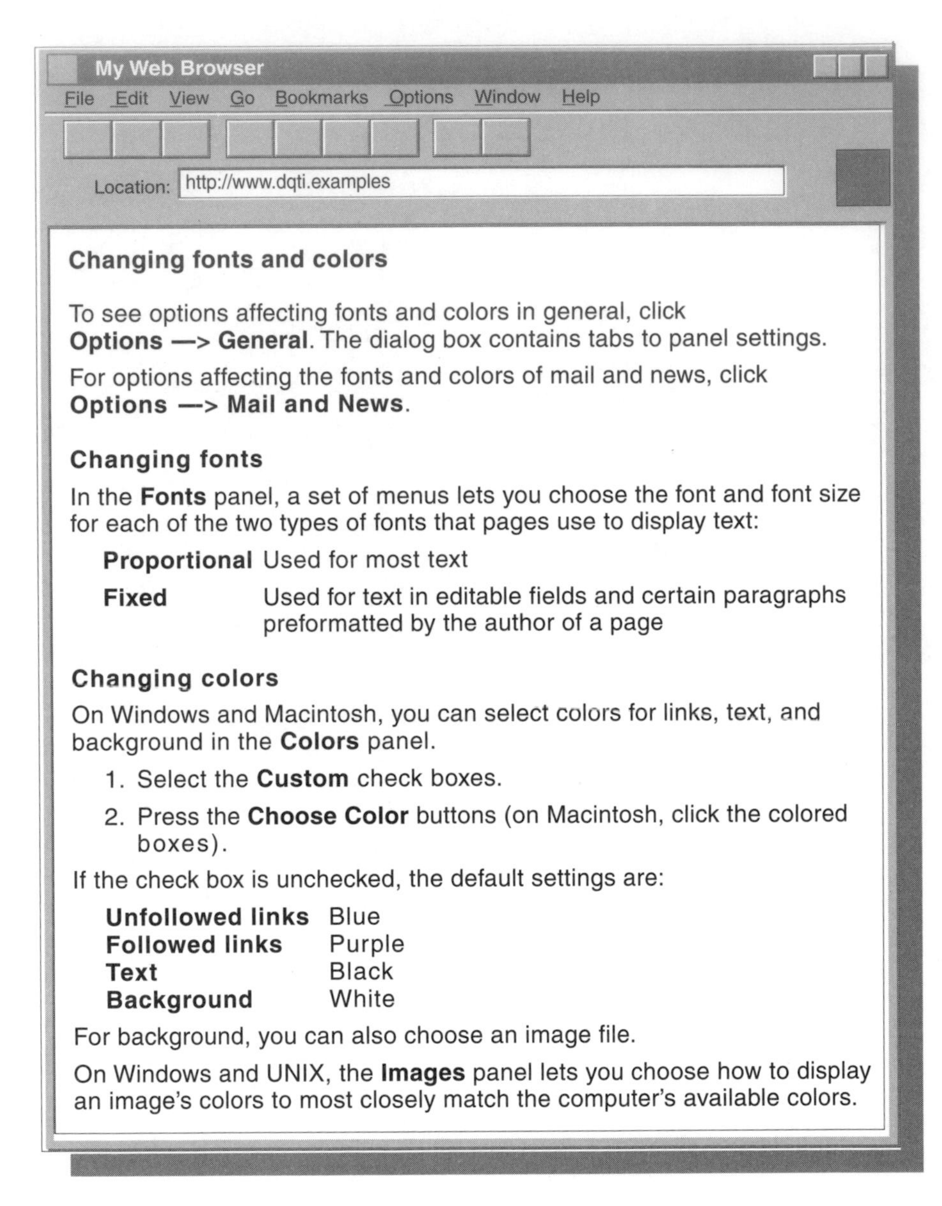

Key words and phrases should stand out to attract attention, especially where they are used for the first time or are defined in the text. After users find a term in the index and go to the topic that it points to, they still need to find the relevant paragraph or sentence. The right kinds of highlighting and structure make this step easier, as shown in the following passages:

Original

Version 4 has the following new features: (1) **long lines**—Version 3 supported up to 255 characters per line, but Version 4 supports up to 1600 bytes (in effect unlimited); (2) **drag and drop** of files and marked text between edit windows and between edit windows and desktop; (3) **fast sort**—much faster than Version 3, with no limit on the amount that can be sorted; (4) **extended grep** adds alternation, grouping, and macros to the standard grep search.

Revision

Version 4 has the following new features:

Long lines	Version 3 supports up to 255 characters per line. Version 4 supports lines up to 1600 bytes, but in effect supports unlimited length lines.
Drag and drop	You can now move files and marked text between edit windows, or between edit windows and desktop.
Fast sort	Sort is much faster than in Version 3, with no limit on the amount that can be sorted.
Extended grep	Alternation, grouping, and macros have been added to the standard grep search.

In the original text, the run-in list makes a specific feature (such as fast sort) hard to find, even with the highlighting. In the revised text, the list is formatted so that users can easily find what they're looking for. Another method of making the same information more retrievable is to put it in a table, with headings to indicate the type of feature and its function or how it has been improved.

Although highlighting is a good way to make key terms stand out, excessive highlighting does more harm than good. Too many highlighted phrases compete for attention and defeat the purpose. And too many different kinds of highlighting (bold, color, italics, underlining) confuse users instead of directing them to key terms and concepts. Compare the following passages:

Original

InfoDBase provides four predefined relationship categories. *Hierarchical* relationship types are used to connect objects that have a hierarchical relationship. *Peer-to-peer* relationship types are used to connect objects that have a peer relationship. *Support* relationship types connect supporting objects to another object (for example, you can connect a **News** object to a **Spreadsheet** object). *Precedence* relationship types connect precedence objects to data resources (for example, you can connect a **Precedence** object to a **File** object). Objects that are connected with this category of relationship are displayed in the Show Lineage Tree window.

The original text shows a paragraph in which several items are highlighted, using two different types of highlighting. The emphasis detracts from the meaning of the passage. Also, the last sentence pertains to only the final object type, but it is the only definition longer than one sentence.

Revision

InfoDBase provides four predefined relationship categories:

hierarchical Connects objects that have a hierarchical relationship.

peer to peer Connects objects that have a peer relationship.

support Connects supporting objects to another object (for example, you can connect a News object to a Spreadsheet object).

precedence Connects precedence objects to data resources (for example, you can connect a Precedence object to a File object). Objects that are connected with this category of relationship are displayed in the Show Lineage Tree window.

The revised passage uses a list and highlighting to emphasize the terms being defined and make their definitions easy to see. The last sentence now clearly belongs with the last term. The object names are no longer highlighted because the use of initial capitals on the names helps to distinguish them from normal words. When you are deciding what types of highlighting to use, consider the effect that highlighting will have on retrievability. The style guideline "Create and follow style guidelines" on page 203 discusses practical and consistent highlighting.

In sum

Retrievability is what ensures that users can find the information that they need, no matter where it is or how it is organized or presented. Without good retrievability, it doesn't matter how accurate or complete your information is—users must find the information to be able to use it.

Use the guidelines in this chapter to ensure that your technical information is retrievable. Refer to the examples in this chapter for practical applications of these guidelines.

When you review technical information for retrievability, you can use the checklist on page 274 in two ways:

- ❑ As a reminder of what to look for, to ensure a thorough review
- ❑ As an evaluation tool, to determine the quality of the information

You can apply the quality rating in the third column of the checklist to the guideline as a whole. Judging by the number and severity of items you found, decide how the information rates on each guideline for this quality characteristic. You can then add your findings to "Quality checklist" on page 387, which covers all the quality characteristics.

Although the guidelines are intended to cover all areas for this quality characteristic, you might find additional items to add to the list for a guideline.

Guidelines for retrievability	Items to look for	Quality rating
Facilitate navigation and search.	• Navigational links are provided. • Meta tags contain the appropriate terms.	1 2 3 4 5
Provide a complete and consistent index.	• Every task, concept, reference topic, and main subject is indexed. • Index entries are specific. • Variations and synonyms for index entries are included. • An online index has only a few variations and synonyms for entries. • A printed index has many variations and synonyms for entries. • A printed index is at least the minimum length. • Entries in a printed index have page numbers on only one level. • Entries in a printed index have no more than two page numbers.	1 2 3 4 5
Use an appropriate level of detail in the table of contents.	• An online table of contents includes all main task topics. • A printed table of contents follows the length guidelines. • If expandable links are used, the highest level clearly indicates what information it contains.	1 2 3 4 5
Provide helpful entry points.	• Paragraphs contain three to five sentences. • Headings, labels, and lists balance the text. • Headings clearly identify the topic, with distinguishing information at the start. • The number of headings is appropriate.	1 2 3 4 5
Link appropriately.	• Links are provided only when helpful. • Links are provided consistently. • Links do not clutter a topic. • In printed information, cross-references in text are kept to a minimum. • There is no more than one inline link for every two paragraphs of online text.	1 2 3 4 5

Guidelines for retrievability	Items to look for	Quality rating
Design helpful links.	• Wording of links and cross-references describes the target information. • Lists of links emphasize the words that are different. • Linking to different locations or levels is minimal. • Redundant links are used only where appropriate.	1 2 3 4 5
Make linked-to information easy to find in the target topic.	• Linked-to information is highlighted or in the first sentence. • Key terms and definitions are appropriately highlighted. • Highlighting of key terms and definitions is appropriate. • Only two or three kinds of highlighting are used.	1 2 3 4 5

Note: The scale for the quality rating goes from very satisfied (1) to very dissatisfied (5).

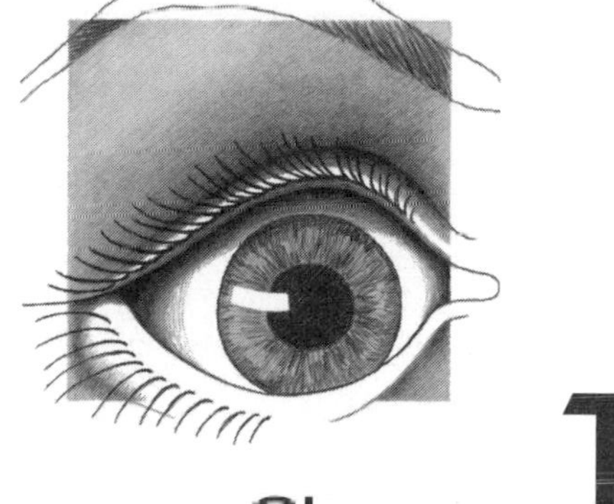

Chapter 10

Visual effectiveness

Art does not reproduce the invisible; rather it makes it visible. —Paul Klee

Visual effectiveness is a measure of how the appearance of information and the use of visual elements within it affect the ease with which users can find, understand, and use the information. Because most of your users use their eyes to access your information, the visual impact of that information can be an important factor in its general quality.

While you are planning and writing your information, you must think of ways to help your information come alive for your users. The guidelines and discussion in this chapter can help you as you work with your information. If are not a "visual" person and you have access to the services of an illustrator or visual designer, enlist those services to help make the invisible visible. If you don't have such services available, use this chapter to help you improve the visual effectiveness of your information.

The visual effectiveness of technical information influences, to a greater or lesser degree, all of the other quality characteristics, and can help them in the following ways:

- ❑ Emphasize the sequence of task steps.
- ❑ Correctly depict facts and their relationships.
- ❑ Make organization more obvious.
- ❑ Complete a description or explanation.
- ❑ Clarify information that might otherwise be confusing or too complex.
- ❑ Make facts "come alive."
- ❑ Set the tone and reinforce a consistent style.
- ❑ Make important information easier to find.

The ever-increasing power, accessibility, and connectivity of computers greatly increases the variety of information delivery media available. Most of these media offer a richness of graphical capabilities that are not available in the print media:

- ❑ Bright, modifiable interfaces and color palettes
- ❑ Faster, flexible Web browsers
- ❑ Easier-to-use software for creating and manipulating images
- ❑ Integration of images, animation, video, and sound with text

Such rich and varied graphical media and methods offer potential for either enhancing or diminishing the quality of technical information. Your goal for visual presentation should be to attract and encourage users to access your information, and then to help motivate them to continue reading. Users are motivated when they perceive a reasonable chance of success at their task. Effective use of visual elements can help users perceive that chance of success and make their search for information more interesting and efficient, and maybe even fun.

To make information visually effective, follow these guidelines:

- ❑ **Use graphics that are meaningful and appropriate.**
- ❑ **Choose graphics that complement the text.**
- ❑ **Use visual elements for emphasis.**
- ❑ **Use visual elements logically and consistently.**
- ❑ **Balance the number and placement of visual elements.**
- ❑ **Use visual cues to help users find what they need.**
- ❑ **Ensure that textual elements are legible.**
- ❑ **Use color and shading discreetly and appropriately.**
- ❑ **Ensure that all users can access the information.**

Use graphics that are meaningful and appropriate

Effective graphics are an integral part of technical information. They can supplement descriptive text to clarify and enliven it. However, graphics that are merely decorative and are unrelated to the text distract and confuse users and can impair their understanding of the information.

Ensure that each illustration accurately depicts the object, concept, or function that it is designed to illustrate and that it does so as simply as possible. Users naturally look at parts of the window or page that contain something other than text. To ensure that your graphics offer meaningful and relevant information you can follow these guidelines:

- ❑ Illustrate significant concepts.
- ❑ Avoid illustrating what is already visible.

Illustrate significant concepts

Carefully choose what you illustrate. Fundamental and significant concepts probably offer the best opportunity for using meaningful graphics. The following passage about features in a new release of a product clearly states facts that might be difficult for users to conceptualize:

Original

> **More data in one table space:** A table space can hold 1 terabyte (TB) of data, instead of only 64 gigabytes (GB).
>
> **More and larger partitions:** Large table spaces can have a maximum of 254 partitions of 4 GB each, instead of only 64 partitions of 1 GB each.

The original passage accurately describes some improvements to a software product. The passage compares some very large numbers to some even larger numbers, but the words alone don't convey the magnitude of the enhancement, or its significance to the users. An illustration helps to make the differences meaningful.

Revision

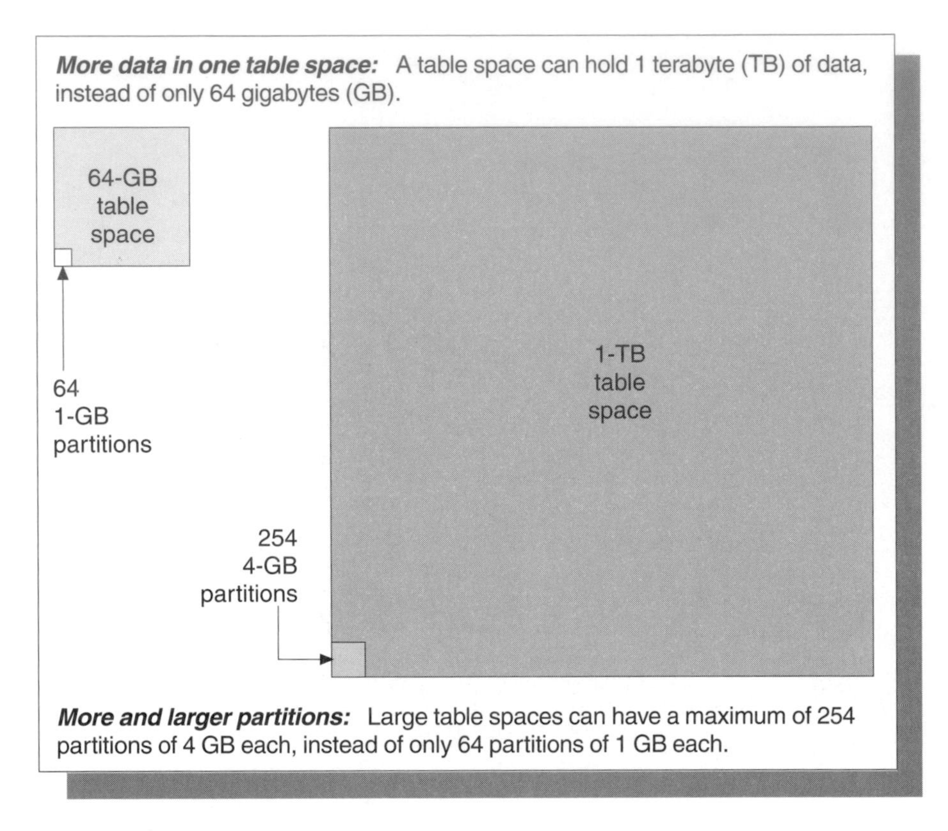

In the revised passage, simple but accurately scaled boxes dramatically compare the capacities in the new release with those in the previous release. Adding this illustration brings the meaning of the words to life.

Avoid illustrating what is already visible

You might be tempted to illustrate the online product interfaces that your text describes. Such captured windows or screens can add visual interest to what might otherwise seem visually uninteresting. Several arguments against using captured windows follow:

- Most importantly, if the product's interface is well designed, you should not need to show a copy of it to explain how to use it. The best interfaces are those that you need not explain at all.

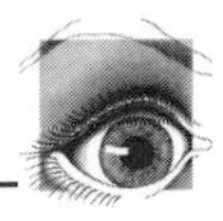

- You cannot always ensure that the captured window in your information reflects the current level of the product. If the image does not match the window that users see, the discrepancy might confuse your users. This recommendation is related to the accuracy guideline "Keep up with technical changes" on page 53.
- If the product and the information are translated, the window needs to be recaptured from the translated product. Translators might not have that capability. Even if they do, the result is an increase in the cost of translation and the time that is needed for translation and testing.
- If users can see the window in the product, why reproduce it? If they must compare the captured window with the actual window, their attention is divided.
- Adding captured windows increases the use of computer memory and storage, the number of pages, and the overall cost.

In some cases the product's visual elements (such as icons, buttons, entry fields) might vary in design or placement from one product window to another. In these cases you might feel that you need to show the product screens. However, a better solution might be to work with the product team to improve the interface rather than illustrate its inconsistencies. When you do need to show product windows, follow these guidelines:

- Ensure that showing the windows adds to users' understanding of the material.

  Usability testing is a good way to determine whether users will benefit from seeing a window image in the information about it.
- Capture all of the windows or window fragments at the same scale so that the text and graphical elements within the windows display at a consistent size.

  The text in the captured windows should appear equivalent to about 8-point type. If some product windows are much larger than others, resize the larger windows, if possible, before capturing them.
- Show only the portion of the interface on which the user must focus to do a task or to find critical information.

  If you show only a portion of a window, however, include the nearest corner of the window. Some users prefer to see complete windows rather than fragments of windows, so you can help these users better understand the physical context of the window fragment by showing them where in the window the fragment occurs.

In the following set of instructions, a screen capture shows users where to find the **Open** menu item:

Original

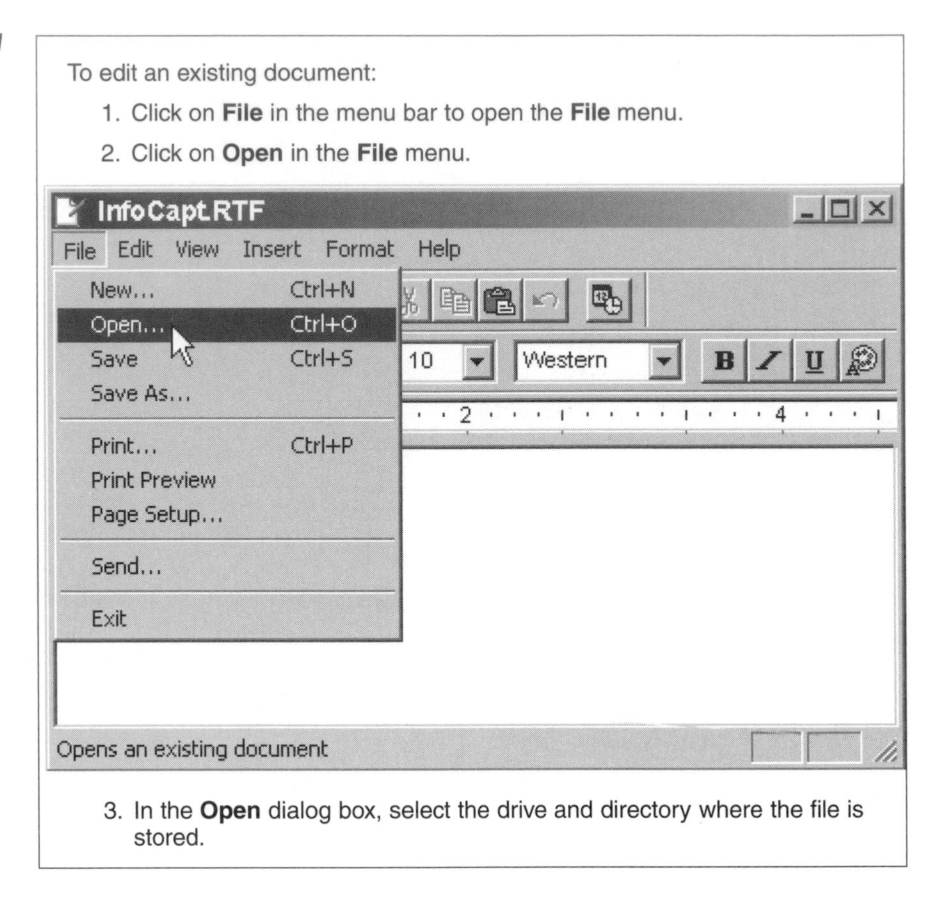

To edit an existing document:

1. Click on **File** in the menu bar to open the **File** menu.
2. Click on **Open** in the **File** menu.

3. In the **Open** dialog box, select the drive and directory where the file is stored.

The original passage shows the entire window of the editing program, with the **File** menu opened and the **Open** item selected. Because the window is so large, however, users cannot easily focus on the section of the window that is relevant to their task. Additionally, depending on how users get to this point, the window on the screen might appear somewhat different from the captured window. For the **Open** task, these differences are probably not important; what is important is selecting **Open** from the **File** menu.

Revision

To edit an existing document:

1. Click on **File** in the menu bar to open the **File** menu.
2. Click on **Open** in the **File** menu.

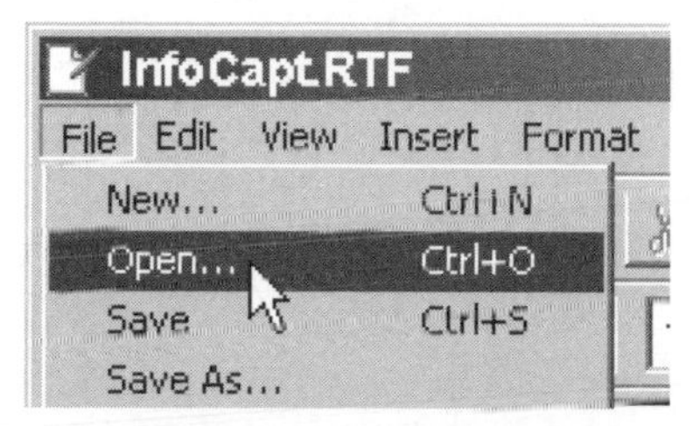

3. In the **Open** dialog box, select the drive and directory where the file is stored.

The revised example shows only the upper left corner of the window, with the **File** menu opened and revealing the selected **Open** action.

Users can more easily recognize both where the **File** menu appears on the window and where **Open** is in the menu. Users are not distracted or confused by extra information that they do not need. By removing those parts of the window that might be distracting or confusing to users, as in the revised passage, you can help them to more quickly and easily focus on the important information.

You can make your users' jobs easier by showing them only the graphics and images that are appropriate to help them understand the information.

Choose graphics that complement the text

Use graphics that help to complete and clarify the information. Relate illustrations to what the user wants to know.

Some users are word oriented; others are picture oriented. Showing as well as telling is an important way to ensure that both types of users get the point. You can show what is unwieldy to tell, highlight what is especially important, and provide clear symbols that users can remember. Telling as well as showing ensures that the information is accessible to visually impaired users, too.

Do not try to replace thorough and accurate text with graphics, however. Although most of your users are likely to access your information visually, some users might not be able to view graphics, so the information in graphics must be available to them in a textual format as well.

Ensure that graphics are neither too elementary nor too complex for the users. Illustrate the information that is most important for users to know, and show it as clearly as possible.

The following wizard contains a brief description of the tasks that the wizard does and four buttons that indicate each task:

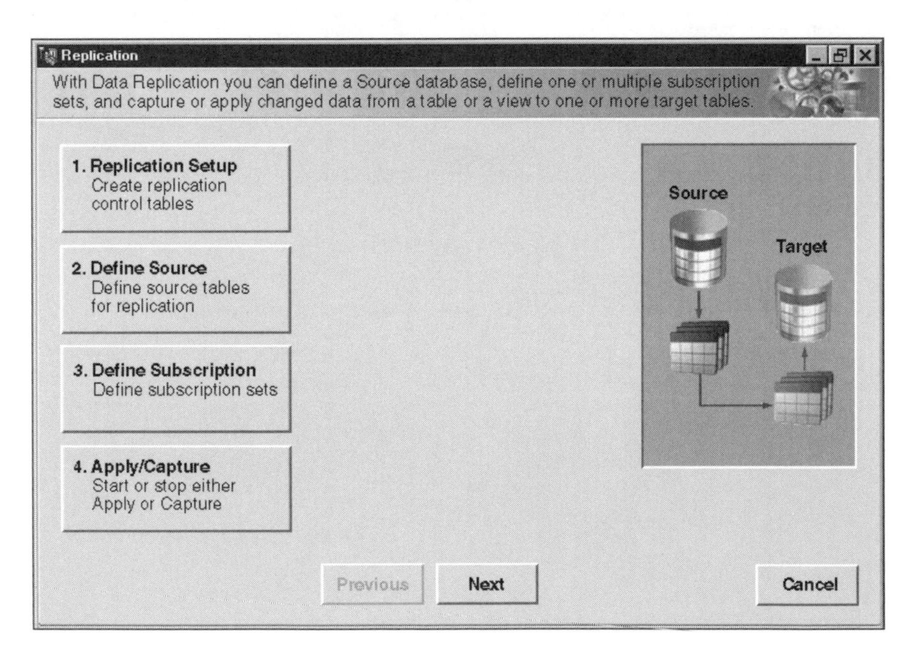

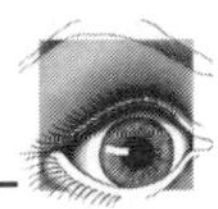

The original graphic is an overly simplified representation of the concepts behind the tasks that the wizard supports. It does not offer any insight into how the four tasks relate to each other, or into how users should interact with the wizard.

Because the wizard supports several tasks, users might appreciate more information about each task than the original wizard offers.

Revision

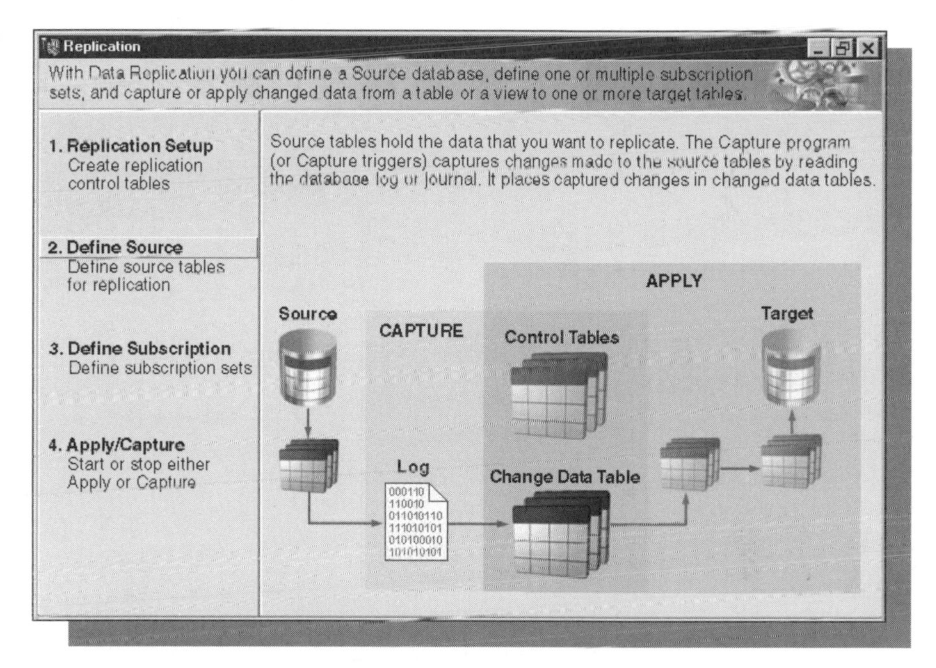

The revised wizard provides a much more detailed graphic that offers a visual overview of the concepts behind each of the tasks and how the four tasks fit together. In addition, as the cursor rolls over each task on the left, a different section of the graphic is highlighted to show how that task fits into the overall process. Descriptive text above the graphic also changes as the graphic changes, to further clarify what the user can accomplish during that task.

So before users even begin to use the wizard, they have a clear understanding of what the wizard is helping them to do. With the revision, the graphic is transformed from a decorative element to one that complements the text and is integrated with it to help users understand the tasks that the wizard supports.

The following passage describes and illustrates a new database software function from the program developer's viewpoint:

Original

> The following illustration represents what takes place during online reorganization. A shadow copy is made of the table spaces. The log is applied to the shadow copy incrementally. When InfoDBase is ready to apply the last increment, writers are drained. After applying the log, readers are drained and all users are switched to the new copy.
>
> Both data and indexes are reorganized at the same time.

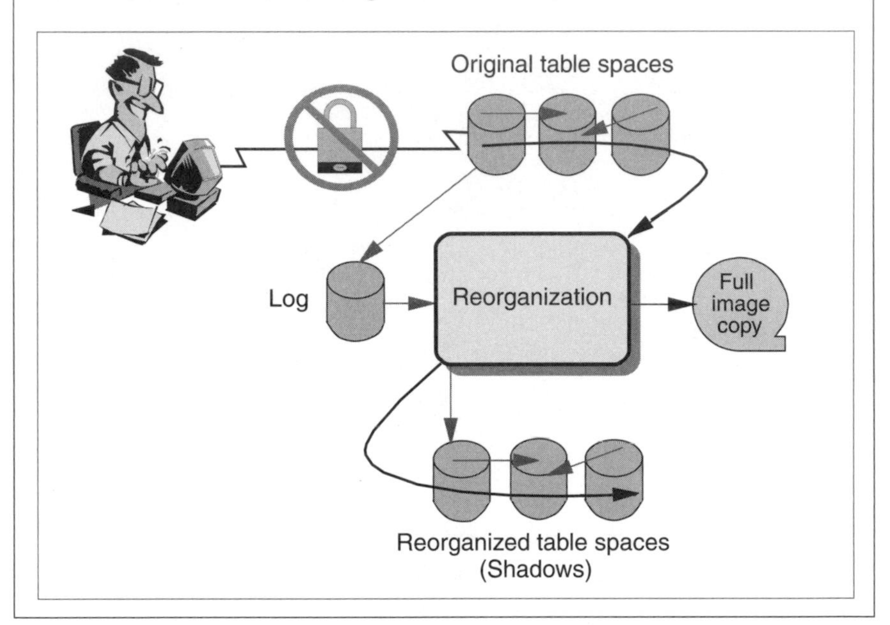

The text that precedes the illustration describes how the database is designed to work. It does not indicate why the new function might be important to users or how they might use it. The original illustration shows several arrows to indicate movement of data during the reorganization of an InfoDBase table space, but the text doesn't adequately explain what the arrows mean or how they differ. The information that is significant to users is that they have write access to their information for a longer period of time with the new function, but the original figure and text neither say nor show this important fact. The cartoon figure of a Caucasian male user adds some levity and interest but no meaning, and is therefore just a distraction. In addition, the figure is neither gender-neutral nor culture-neutral. For information about making your information gender-neutral and culture-neutral, see the style guideline "Use unbiased language" on page 208.

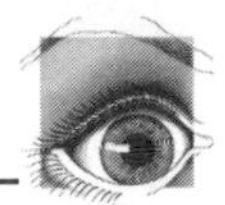

Revision

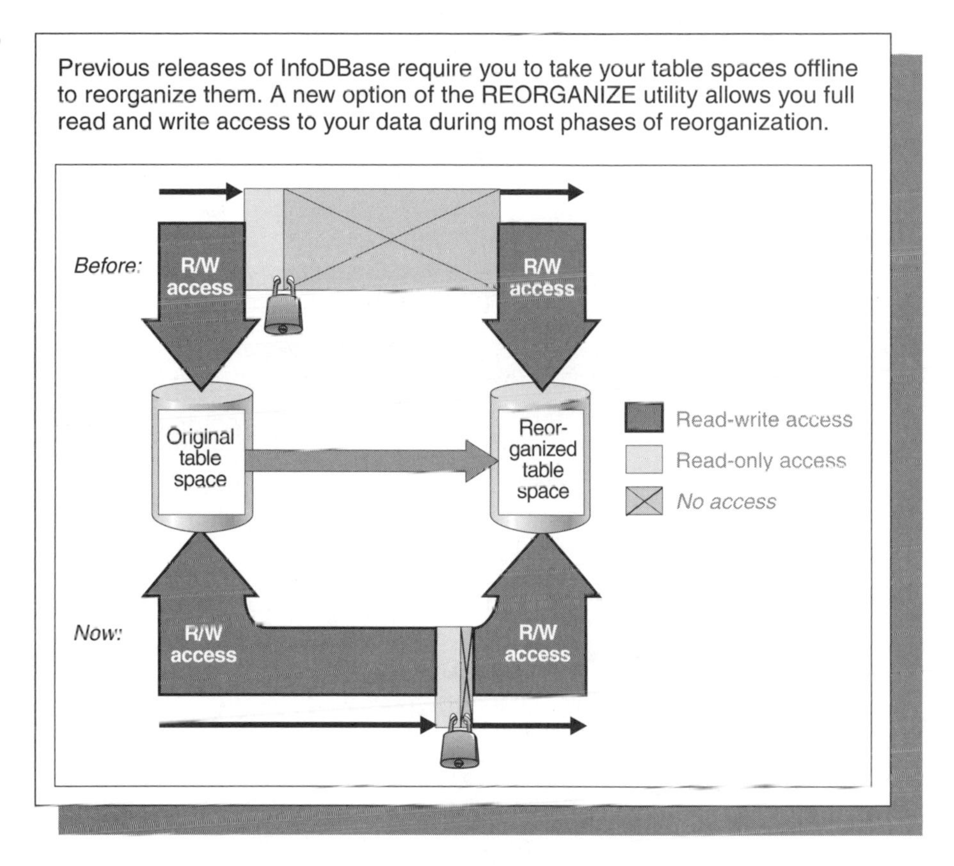

Previous releases of InfoDBase require you to take your table spaces offline to reorganize them. A new option of the REORGANIZE utility allows you full read and write access to your data during most phases of reorganization.

The revised illustration shows how the new function is significant to users. Users want to know why the reorganization function is better in the new release than in the previous release of the product. The revised illustration shows a comparison of the old function and the new, emphasizing the difference in read-write access availability during the reorganization process. Very little text is needed to tell the user that the new function is an improvement over the old.

When you add illustrations to your information, you must ensure that they are meaningful to your users and that they both complement and supplement the text. Do not use illustrations as substitutes for clear, accurate, and complete textual information; rather, use them to make your information more concrete.

Use visual elements for emphasis

Visually emphasize the most important information. Visual elements can help make the organization of the information more obvious; they can also help to focus the user's attention on the most important information.

Highlighting words or phrases gives them greater importance by making them stand out visually. You can increase this effect with visual elements such as symbols and icons, rules, bullets, tabs, color, and shading.

If a phrase or paragraph is especially important, highlight it with a font change, or place it in a box or other structure to set it apart from the text that surrounds it. In an illustration, if one object or portion of the illustration is particularly significant, emphasize it with a slightly heavier outline, shading or color, or highlighted label. If necessary, try different techniques and then conduct an informal usability test to see which method best highlights the most important information. Sometimes, for example, adding shading to an object makes it recede rather than stand out, while adding shading to the background and leaving the object unfilled emphasizes it.

Whatever methods you choose, use them consistently for similar emphasis throughout the information. Follow these guidelines when using visual elements for emphasis:

- ❑ Emphasize the appropriate information.
- ❑ Ensure that your visual elements are not distracting.

Emphasize the appropriate information

Effective use of visual elements for emphasis can be difficult. You can emphasize text by changing the font, but in an illustration where text and graphical objects work together, you might need to emphasize objects as well as text. You might find that a technique that emphasizes an object obscures the text within it. A color that emphasizes an object when viewed in color might actually de-emphasize that object when viewed in black and white, as might happen when a user prints a PDF file on a black and white printer. See "Ensure that the contrast between text and background is adequate" on page 316 for more discussion on the use of color.

The following illustration shows how an improvement in a product function enables users to merge information from two tables in several ways:

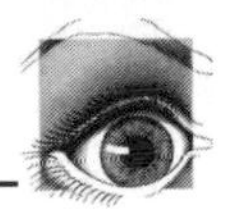

Original

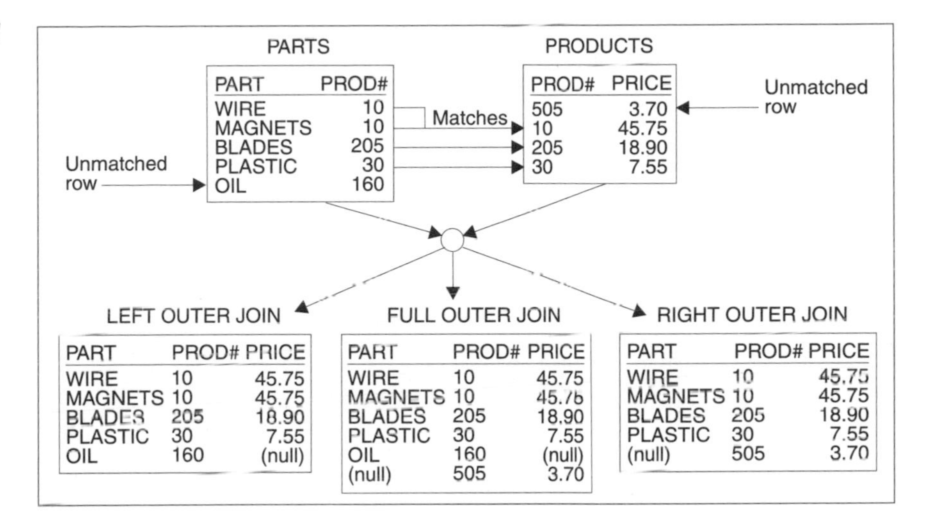

With the original illustration, users must read the entire contents of each box to understand the relationships among them and thereby understand the meaning of the illustration. Consider how you might use shading to make the illustration more effective:

First revision

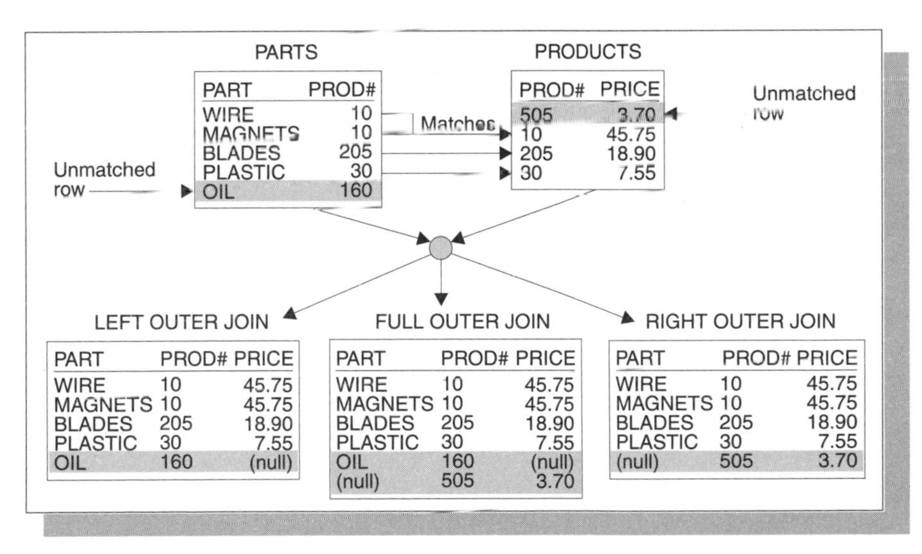

When shading is added to certain lines of text in the illustration, as in the first revision, those lines are emphasized. Users can more easily identify the significant lines in each box and thus the relationships between them. Users can more easily grasp the point of the illustration.

But if the shading is too dark or too coarse, the text in those lines can become more difficult to read than the unshaded text. So how can you emphasize those lines without also making them less distinct? You might try a technique that is similar to putting a spotlight on the significant lines:

Second revision

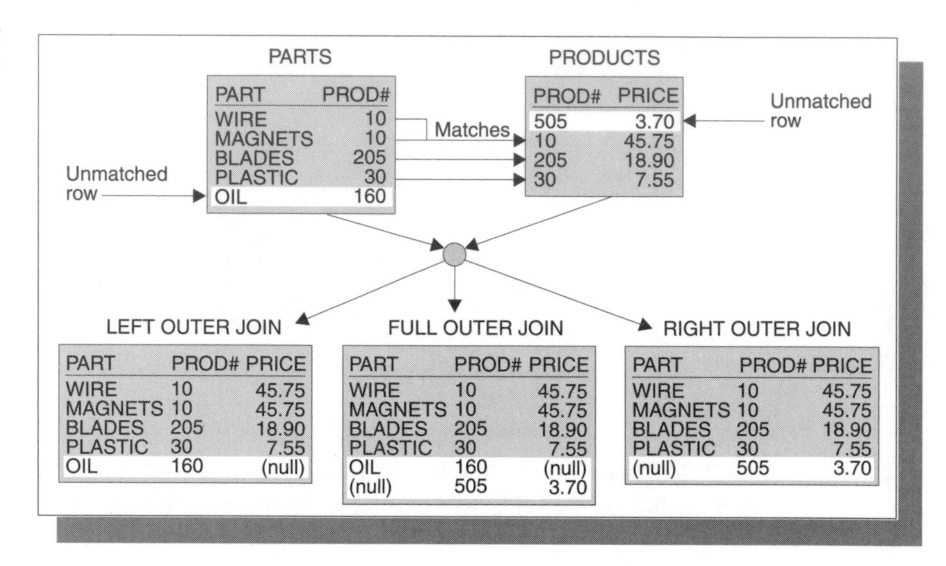

When shading is added to all but one or two lines of text in the boxes, as shown in the second revision, the unshaded parts are emphasized. At the same time, the less important lines are slightly less distinct. Users can more easily see the unshaded lines and their relationships to one another.

You might want to show how some information appears in the context where users will find it. If the context is busy and complex, users might have difficulty focusing on the important part.

The following window appears in a tutorial to help users understand how to use a software product interface:

Original

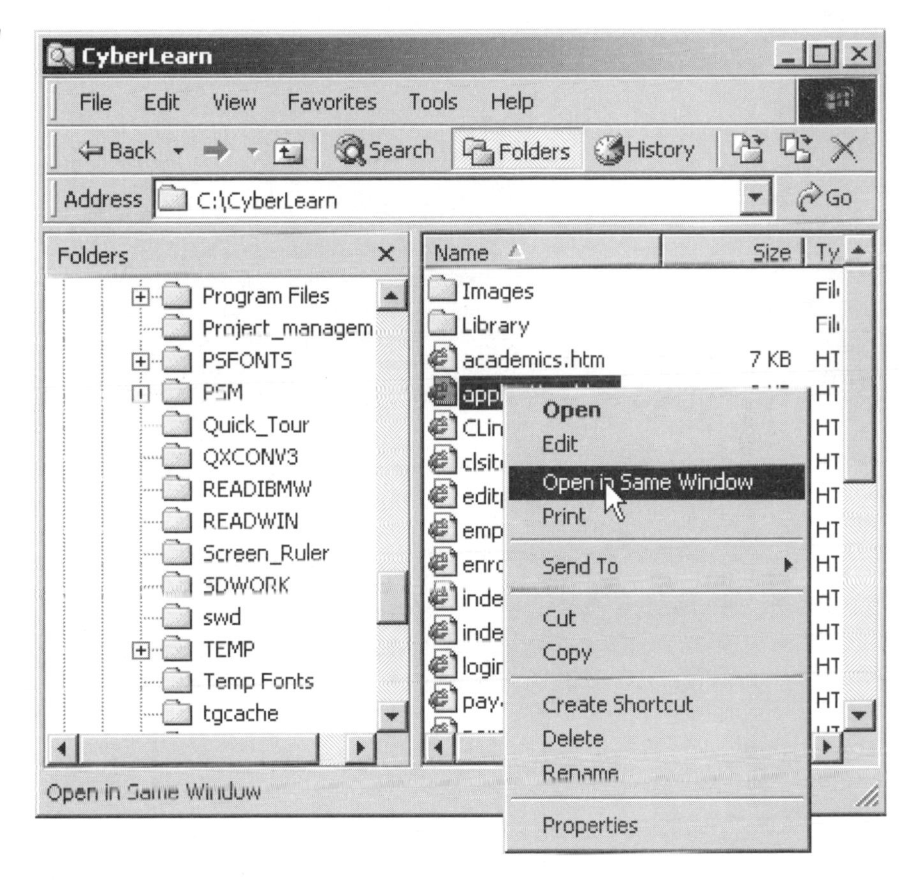

In the original graphic, the complete window is shown along with the menu that is the focus of attention in the tutorial. Although the menu is obvious, it is not emphasized in a way to draw users' immediate focus.

Revision

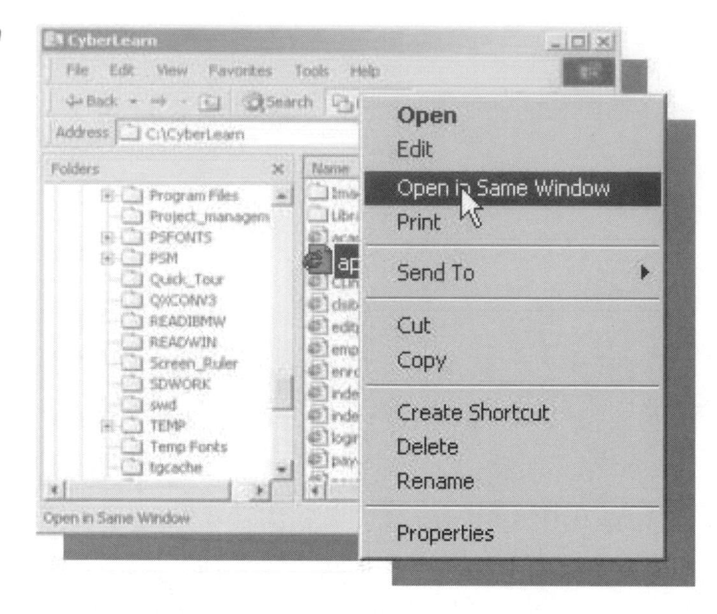

In the revised graphic, the menu is left at full size and intensity, while the rest of the window is reduced in size and contrast. The resulting graphic shows the entire window but de-emphasizes everything except the menu and the file to which it belongs. Users' attention naturally falls on the significant information.

You can help users focus on the significant information in an illustration by both emphasizing the important parts of the graphic and de-emphasizing the less important parts.

Ensure that your visual elements are not distracting

Do not make your visual elements so strong that they become the primary focus on the window or page. The information is what is important; the structures that make it easier to find and understand should recede so that they simply support, but do not overpower, the information.

Consider the use of rules in the following table:

Original

Breed	Origin	Characteristics	Mane and tail type
Quarter horse	Breed developed in Western U.S.	Sprint speed; tight turns	Medium texture and thickness; mane often pulled or trimmed short
Morgan	Breed developed in Eastern U.S.	Versatility; endurance; hauling strength; elegant trot	Thick, long, coarse
Arabian	Ancient breed developed in Arabian desert	Endurance; speed; elegance and flash	Fine and flowing; mane generally kept long
Appaloosa	Technically not a breed; rather, a set of coloring characteristics that occur in several breeds	Two or more of these characteristics ❑ Visible white sclera ❑ Sparse mane and tail ❑ Striped hooves ❑ Spotted skin	Medium coarse; usually sparse

The original table stands out on the page, but the information within it is overpowered by the table rules, particularly the vertical rules, which make it more difficult for the eyes to travel smoothly across the rows and from one row to the next.

The revised table makes the organization of the information clear without imposing the rules as a barrier to the information flow:

Revision

Breed	Origin	Characteristics	Mane and tail type
Quarter horse	Breed developed in Western U.S.	Sprint speed; tight turns	Medium texture and thickness; mane often pulled or trimmed short
Morgan	Breed developed in Eastern U.S.	Versatility; endurance; hauling strength; elegant trot	Thick, long, coarse
Arabian	Ancient breed developed in Arabian desert	Endurance; speed; elegance and flash	Fine and flowing; mane generally kept long
Appa-loosa	Technically not a breed; rather, a set of coloring charac-teristics occurring in several breeds	Two or more of these characteristics ❑ Visible white sclera ❑ Sparse mane and tail ❑ Striped hooves ❑ Spotted skin	Medium coarse; usually sparse

In the revised table, faint rules separate the rows, and consistently placed white space separates the columns, making their layout clear while allowing users to easily scan and focus on the data. The horizontal rules are just visible enough to help users scan across the row, but it is the text in the table that stands out.

Use graphical techniques to emphasize the important information on a window or page. Ensure that it is the information that stands out rather than the graphical technique or document structure the information is in.

Use visual elements logically and consistently

Choose a visual system to use throughout your information. A visual system specifies consistent characteristics for all of the visual elements in your information. It can address everything from the type style used for emphasis to the weight and placement of rules in tables and lines in illustrations.

When the same visual system is used throughout the information, users quickly learn where and how to find the information that they need. You should have guidelines that apply to headings; highlighting; font usage; examples of code; syntax formats; list, table, and figure formats; indention; and graphic styles and placements. If the style guidelines that you are following (as discussed in "Create and follow style guidelines" on page 203), do not address consistent treatment of visual elements, work with a visual designer to establish a visual system. Then add the standards to your style guidelines, and stick to them.

You are probably working with text-formatting software that applies a standard style to your information through a cascading style sheet or document template. However, style sheets and templates often allow some flexibility for customizing the presentation of such elements as lists, tables, and illustrations. You should have a visual system or supplemental style guidelines for such elements.

If you can create your own styles and formats for your text-formatting software, do so with the advice of an editor and a visual designer. Ensure that your template is visually logical and consistent. To ensure visual consistency, follow these guidelines:

- ❑ Use a visually simple but distinct heading hierarchy.
- ❑ Maintain consistent placement of document elements.
- ❑ Ensure that the look and feel of multimedia presentations is consistent.
- ❑ Use icons and symbols consistently.

Use a visually simple but distinct heading hierarchy

One way to ensure a visually logical and consistent template is to pay close attention to the visual presentation of your heading hierarchy.

Headings organize information into discrete chunks. Differences in size, weight, and placement of headings give users visual cues to the relative positions of the headings in the organizational hierarchy. The visual differences in the headings should be noticeable but not extreme.

If your headings are numbered, their position in the hierarchy is immediately apparent. If headings are not numbered, you can use any of several methods to distinguish them:

- Font
- Point size
- Weight (bold, regular, light)
- Style (normal, italic, underscored, overscored)
- Case (all capitals, headline style, sentence style)
- Indention

Using more than three of these methods, however, is more likely to confuse and distract users than to help them understand the organizational hierarchy.

The following outline presents five heading levels:

Original

Building a Better Web Site

Designing your Web site

Organizing the Web site

DEVELOPING THE SITE MAP

Coding the page

The original outline uses all the differentiation methods except font change to signal a change in heading level:

- Level 1 is 16-point bold, headline style.
- Level 2 is 14-point bold italic, indented (1), sentence style.
- Level 3 is 12-point regular, underscored, indented (2), sentence style.
- Level 4 is 11-point regular, indented (2), all capitals.
- Level 5 is 10-point bold, indented (3), sentence style.

The highlighting methods used seem to have little relationship to one another, so users might have trouble remembering where an individual heading falls in the hierarchy. A more orderly progression of changes, as in the revised outline, is easier to follow:

Revision

> **Building a better Web site**
> **Designing your Web site**
> **Organizing the Web site**
> **Developing the site map**
> ***Coding the page***

The revised outline uses only size, indention, and style to distinguish heading levels, and each change affects only one characteristic:

- Level 1 is 16-point bold.
- Level 2 is 13-point bold.
- Level 3 is 13-point bold, indented (1).
- Level 4 is 11-point bold, indented (1).
- Level 5 is 10-point bold italic, indented (1).

The hierarchy in the revised outline is easy to see and to remember.

When you use only two or three methods to distinguish heading levels, users can more easily see the relationships among the headings and recognize a specific heading level, even if it occurs by itself on a page.

Maintain consistent placement of document elements

Attentive users notice even the smallest change in a visual system and assume that it has significance. If it doesn't, or if you haven't explained the difference, users will be distracted and confused. Consider the significance of the placement of examples of code in the following passage:

Original

1. Decide which module to place in maintenance mode.
2. Stop all modules by entering the following command for each module in the group:
   ```
   -IN1G STOP INFOB MODE(QUIESCE)
   ```
3. Start the surviving member in maintenance mode by using the following command:
   ```
       -IN1G START INFOB ACCESS(MAINT)
   ```
4. Stop the surviving member by using the following command:
   ```
    -IN1G STOP INFOB MODE(QUIESCE)
   ```
5. Stop any IQMs that have not stopped, using the following command:
   ```
           STOP iqmproc
   ```

When some code examples are indented and others are not, as in the original passage, or when they are indented a different number of spaces, users might assume that the spacing is significant, and they might code an application incorrectly.

Revision

1. Decide which module to place in maintenance mode.
2. Stop all modules by entering the following command for each module in the group:
   ```
   -IN1G STOP INFOB MODE(QUIESCE)
   ```
3. Start the surviving member in maintenance mode by using the following command:
   ```
   -IN1G START INFOB ACCESS(MAINT)
   ```
4. Stop the surviving member by using the following command:
   ```
   -IN1G STOP INFOB MODE(QUIESCE)
   ```
5. Stop any IQMs that have not stopped, using the following command:
   ```
   STOP iqmprocSTOP iqmproc
   ```

In the revised passage, all examples of code are aligned with the left margin of the list text. Users can be confident that code that is similar to one of the examples does not need to be indented to work correctly.

Similarly, if every table has a different format, users must learn how to identify and read each table that they encounter. The following two tables, for example, both present information about a product, but in different formats:

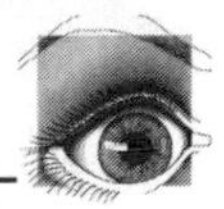

Original

Table 1. Values inserted in the INFODA.

Value	*Field*	*Description*
INFODA	INFODAID	An "eye-catcher"
8816	INFODABC	The size of the INFODA in bytes (16 + 44 × 200)
200	INFON	The number of occurrences of INFO-VAR, set by the program
200	INFOD	The number of occurrences of INFOVAR actually used by the DESCRIBE statement
452	INFOTYPE	The value of INFOTYPE in the first occurrence of INFOVAR. It indicates that the first column contains fixed-length character strings and does not allow null values.

INFO data type	**InfoDBasic equivalent**	**Examples**
SMALLINT	INTEGER*2	
INTEGER	INTEGER*4	
DECIMAL(*p,s*) or NUMERIC(*p,s*)	no exact equivalent; use REAL*8	
FLOAT(*n*) single precision	REAL*4	1<*n*<21
FLOAT(*n*) double precision	REAL*8	22<*n*<53
CHAR(*n*)	CHARACTER**n*	1<*n*<254

Table 2. INFO data types mapped to typical InfoDBasic declarations.

The first table has its caption at the top, it has no vertical rules, and its column headings are italicized. The caption for the second table is at the bottom, the table contains vertical rules, and its column headings are bold.

Before users can extract the important information from the second table in the original passage, they must identify how the table differs from the earlier one and then determine whether the differences have meaning. The inconsistent table format has slowed their progress and comprehension needlessly.

Revision

Table 1. Values inserted in the INFODA.

Value	Field	Description
INFODA	INFODAID	An "eye-catcher"
8816	INFODABC	The size of the INFODA in bytes (16 + 44 × 200)
200	INFON	The number of occurrences of INFO-VAR, set by the program
200	INFOD	The number of occurrences of INFOVAR actually used by the DESCRIBE statement
452	INFOTYPE	The value of INFOTYPE in the first occurrence of INFOVAR. It indicates that the first column contains fixed-length character strings and does not allow null values.

Table 2. INFO data types mapped to typical InfoDBasic declarations.

INFO Data Type	InfoDBasic Equivalent	Examples
SMALLINT	INTEGER*2	
INTEGER	INTEGER*4	
DECIMAL(*p,s*) or NUMERIC(*p,s*)	no exact equivalent; use REAL*8	
FLOAT(*n*) single precision	REAL*4	$1<n<21$
FLOAT(*n*) double precision	REAL*8	$22<n<53$
CHAR(*n*)	CHARACTER**n*	$1<n<254$

When table structures are consistent, as in the revised passage, users quickly learn to use the structure only as a guide and to focus on the information that the table contains.

Consistent use and placement of document elements and structures help users learn quickly how information is presented. Familiar document structures almost disappear, leaving users free to focus only on the information.

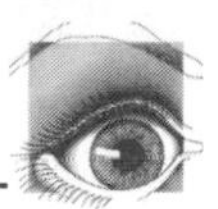

Ensure that the look and feel of multimedia presentations is consistent

Similar consistency guidelines apply to the design of multimedia presentations, demonstrations, and tutorials. Most software for the devlopment of multimedia offers an almost unlimited selection of window backgrounds, transitional effects, and ability to integrate audio and video clips. Trying all the options might seem exciting, but resist the urge to use all the graphic gimmicks at your disposal. To help users focus on the information rather than the presentation techniques, follow these guidelines:

- Select a screen background that complements your presentation but doesn't overpower it, and use the same background throughout the presentation.
- If your presentation uses dynamic transitional effects such as "fades," choose one or two types and use them judiciously and consistently.
- Use consistent navigation and button placement throughout the presentation.

The following two windows appear in the same presentation:

Original

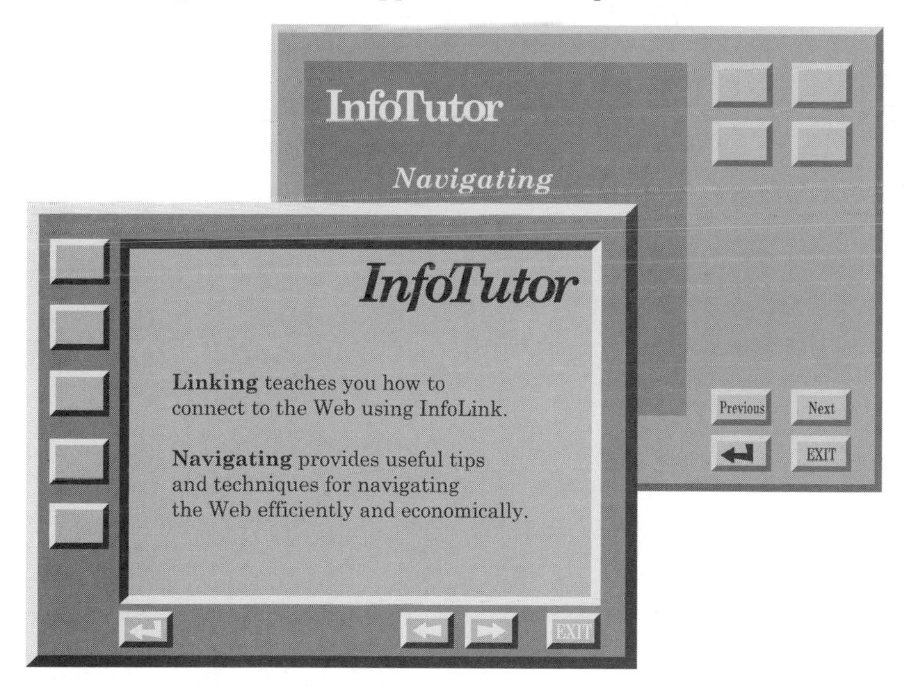

In the first window, contents buttons are on the left and navigation buttons are along the bottom of the window. In the "Navigating" window, however, the contents buttons appear on the upper right, with all the navigation buttons on the lower right. Will users find the original pair of windows easy to use?

Users must first determine, for each of the original windows, which parts of the window contain significant information. Then they must find and understand how to move from that window to another one.

Revision

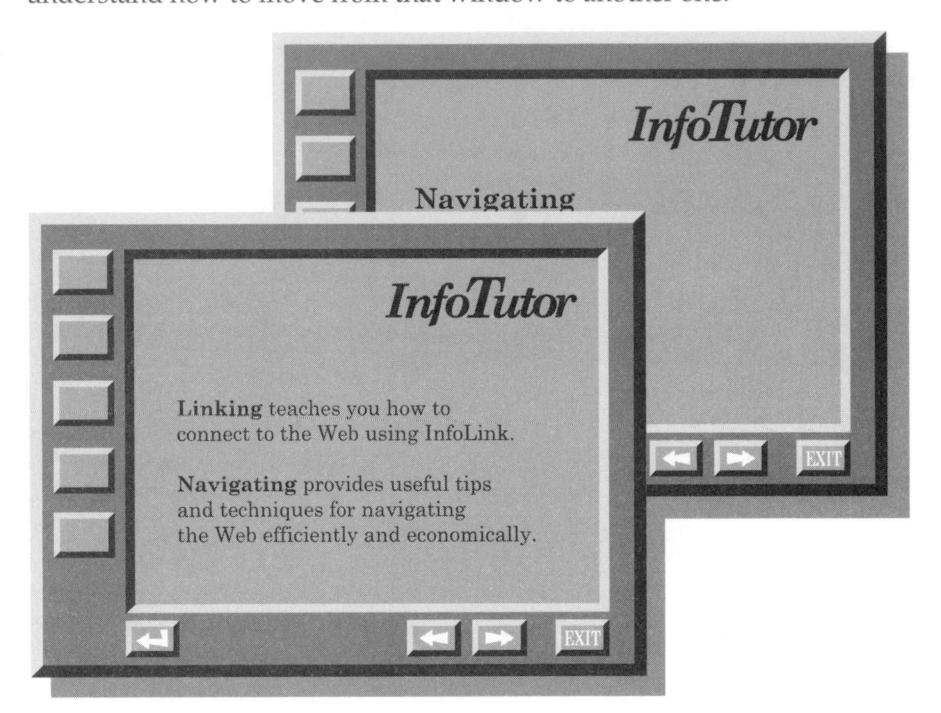

When the presentation methods are consistent, as in the revised windows, users quickly learn the location of the visual elements in the window. Users then notice these elements only as enhancements to the information and, when they move to the next window, they can focus on the information rather than on relearning the presentation and navigation methods.

Just as you should keep document elements and structures consistent in design and placement, so should you keep multimedia presentations consistent. Users need to be able to find and understand the information you present without needing to relearn where to look and how to navigate when a new page appears.

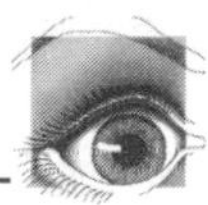

Use icons and symbols consistently

Even the smallest inconsistencies can confuse users. Consider the following buttons, both from the same Web application. Users will most likely assume that both are buttons because their design is identical.

Original

Both buttons look like buttons, and as the cursor rolls over the **Subscribe** button, the cursor changes to a hand, to reinforce users' belief that, when they click the button, a window will open to enable them to enter subscription information. The cursor does not change when it rolls over the **Search** button. Does this mean that the search function does not work? Most users will click anyway, but they might be confused or frustrated at having to test a control to be sure that it will do what they expect it to do.

Revision

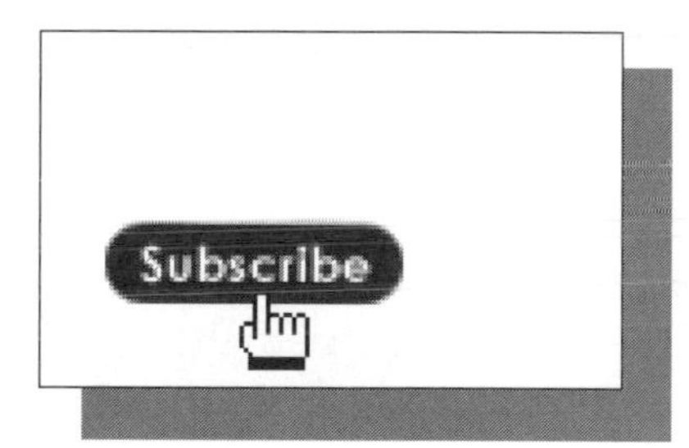

The revised example shows that the cursor changes to the hand when it rolls over both buttons. The consistent use of the hand reinforces users' understanding both that a rounded rectangle that contains text is a button and that the cursor will change to indicate a control that users can click.

Logical and consistent use of visual elements can help users feel comfortable and confident while using your information. When you use a consistent visual system, users can learn it quickly and then focus their attention on the information. The visual elements become so familiar to users that the elements almost merge with the background, supporting the information and the users' journey through it.

Balance the number and placement of visual elements

Everything that the user can see on the window or page contributes to, or detracts from, the visual effectiveness of the information:

- Layout
- Color and shading of background, text, and graphics
- Contrast between background and text
- Typeface and type size
- Illustrations, icons, and animations
- Tables, charts, and code examples
- Navigational cues
- Length of lines and paragraphs
- Highlighting
- Headings
- List styles and lengths
- Overall information density and white space

Users are easily overwhelmed and discouraged by large blocks of text in technical information. By using a variety of visual elements, you can help users find information easily and confidently. When paragraphs are broken up by other visual elements, users are able to pause, evaluate, and absorb the information before moving on. Such elements also make information more visually accessible, interesting, and memorable. This guideline is related to the retrievability guideline "Provide helpful entry points" on page 257.

White space can also offer a place for the eyes to rest and can make a window, page, or two-page spread appear less dense and intimidating. Be sure, though, that extra white space does not interrupt the flow of information on the window or page; such an interruption might suggest to the user that something is missing.

The original page that the following passage represents has many quality problems. This passage is displayed in a Greek font to allow you to focus exclusively on the visual aspects of the page. The page contains both reference and task information, but these are not differentiated in any way. The dense and unvarying presentation of the information impedes rather than helps the user.

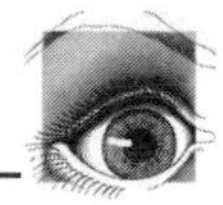

Original

Τηε μοστ χομμον στοραγε σηορταγε προβλεμσ ινϖολϖε Κ9Η. ΜΧΑΤ υσεσ λαργε αμουντσ οφ Κ9Η φορ ιτσ χοντρολ βλοχκσ ανδ βυφφερσ χονταινινγ δατα τηατ ισ βεινγ σεντ αρουνδ τηε νετωορκ. Τηισ προβλεμ χαν βε δυε το α ϖαριετψ οφ χαυσεσ. Φορ εξαμπλε, ιφ ονε αππλιχατιον ισ φλοοδινγ ανοτηερ ωιτη δατα, ανδ τηε σεχονδ αππλιχατιον ισ υναβλε το ρεχειϖε τηε δατα ανδ προχεσσ ιτ ατ αν αδεθυατε ρατε, τηεν τηε βυφφερσ βυιλδ υπ ιν ΜΧΑΤ στοραγε. Τηισ σιτυατιον χαν βε αϖοιδεδ βψ υσινγ σεσσιον παχινγ.

Ιφ ΜΧΑΤ ισ υναβλε το γετ ενουγη στοραγε το ισσυε α μεσσαγε, τηεν τηε νορμαλ στοραγε σηορταγε μεσσαγε ισ αχχομπανιεδ βψ αν ΙΣΤ999Ε μεσσαγε.

ΦΓΗΜΠ ανδ ΘΦΡΔΤ: Τηε Πριμαρψ ρετυρν χοδε (ΦΓΗΜΠ) ανδ σεχονδαρψ ρετυρν χοδε (ΘΦΡΔΤ) αρε βοτη γιϖεν το ΔΔΦ ωηεν αν ΑΠΠΧΧΜΔ μαχρο ηασ χομπλετεδ. Ον οχχασιον, τηεσε χοδεσ μαψ ινδιχατε τηατ τηερε ισ α στοραγε σηορταγε ιν ΜΧΑΤ. Φορ εξαμπλε, ιφ ΦΓΗΜΠ ισ ξ□0037□ ανδ ΘΦΡΔΤ ισ ξ□0000□, τηεν τηισ ινδιχατεσ α στοραγε σηορταγε ωηιλε ΜΧΑΤ ωασ ρεχειϖινγ δατα ορ σενδινγ α παχινγ ρεσπονσε. Α ΦΓΗΜΠ οφ ξ□0093□ ωιτη ΘΦΡΔΤ οφ ξ□0000□ ινδιχατεσ τηερε ισ α τεμποραρψ στοραγε σηορταγε ωηιλε σενδινγ δατα. Υσυαλλψ τηισ ρετυρν χοδε μεανσ τηατ τηε σενδ ρεθυεστ ηασ τεμποραριλψ δεπλετεδ τηε βυφφερ ποολ το συχη αν εξτεντ τηατ τηε ποολ μυστ βε εξπανδεδ. Τηε εξπανσιον ηαδ νοτ οχχυρρεδ βεφορε τηε χομπλετιον οφ τηε ΑΠΠΧΧΜΔ μαχρο.

ΣΝΑ σενσε χοδε: Σομε ΣΝΑ σενσε χοδεσ ινδιχατε τηερε μαψ βε α στοραγε σηορταγε αλσο. Φορ εξαμπλε, α υσερ μιγητ βε αχχεσσινγ δατα φρομ τηε ρεμοτε δαταβασε, ανδ ρεχειϖε □037Χ0000□. Χηεχκινγ τηισ σενσε χοδε ιν τηε μανυαλ ινδιχατεσ τηατ τηερε ισ α περμανεντ ινσυφφιχιεντ ρεσουρχε χονδιτιον. Τηισ ρεσουρχε χουλδ βε στοραγε. Οτηερ σενσε χοδεσ, συχη ασ □300Α000□ ινδιχατε α στοραγε τψπε προβλεμ, βυτ νοτ νεχεσσαριλψ α στοραγε σηορταγε.

Ερρορσ: Τηερε αρε α σεριεσ οφ Π9Λ ερρορσ ωηιχη ινδιχατε στοραγε σηορταγε προβλεμσ. Φορ εξαμπλε, ΕΡΡΟΡ373 ανδ ΕΡΡΟΡ30Α. Τηεσε αρε υνχομμον ιν ΜΧΑΤ.

Ηανγσ: Δεπενδινγ ον τηε στοραγε σηορταγε ανδ τηε προχεσσινγ τηατ ισ οχχυρρινγ, τηε στοραγε σηορταγε χουλδ μανιφεστ ιτσελφ ιν α ηανγ σιτυατιον. Φορ εξαμπλε, ιφ α ϖιρτυαλ ρουτε βεχομεσ βλοχκεδ δυε το στοραγε σηορταγεσ, τηεν αλλ τηε σεσσιονσ τηατ ηαδ βεεν υσινγ τηατ σεσσιον ηανγ υντιλ τηε στοραγε σηορταγε ισ ρελιεϖεδ ανδ τηε ϖιρτυαλ ρουτε βεχομεσ οπεν αγαιν. Ωηεν ανψ οφ τηεσε ινδιχατιονσ οφ α στοραγε προβλεμ ισ ρεχειϖεδ, τηε φολλοωινγ στεπσ χαν βε υσεδ το φινδ ουτ μορε ινφορματιον αβουτ τηε σηορταγε:

Δετερμινε τηε αρεα οφ στοραγε σηορταγε. Τηισ ισ ιμπορταντ, ασ τηε ΜΧΑΤ δισπλαψ χομμανδ ηασ ινφορματιον αβουτ τηε ΜΧΑΤ Κ9Η υσαγε. Τηισ δοεσ νοτ ηελπ διαγνοσισ ιφ τηε στοραγε σηορταγε ισ ιν ΜΧΑΤ πριϖατε. Ηοωεϖερ, περηαπσ Κ9Η υσαγε σηουλδ προβαβλψ βε χηεχκεδ ανψωαψ.

Δισπλαψ βυφφερ υσαγε: Ιφ τηε στοραγε σηορταγε ηασ οχχυρρεδ ιν ονε οφ τηε Κ9ΙΙ συβποολσ, τηεν α □Π ΣΕΤ,ΔΧςΡΦ□ χομμανδ χαν βε υσεδ το δετερμινε τηε αμουντ οφ Κ9Η στοραγε βεινγ υσεδ βψ ΜΧΑΤ. Υσυαλλψ τηε βυφφερ υσε δισπλαψ γιϖεσ α γοοδ δεα οφ ωηιχη ΜΧΑΤ ποολ ισ χαυσινγ τηε στοραγε σηορταγε.

Μονιτορ βυφφερ υσαγε: Ωηεν λοοκινγ ατ α βυφφερ σηορταγε, ιτ ισ οφτεν ηελπφυλ το κνοω ωηετηερ τηε ονσετ οφ τηε προβλεμ ωασ γραδυαλ ορ ιμμεδιατε. Ιφ ρεγυλαρ βυφφερ υσαγε δισπλαψσ αρε δονε, τηεν γραδυαλ ινχρεασεσ ιν βυφφερ υσε χαν βε σεεν. Τηεσε γραδυαλ ινχρεασεσ μαψ τακε δαψσ το μανιφεστ τηεμσελϖεσ ιντο στοραγε σηορταγε προβλεμσ. Ιν φαχτ, ιφ ΜΧΑΤ ισ τακεν δοων ρεγυλαρλψ, τηε στοραγε σηορταγε σψμπτομ μαψ νεϖερ βεεν σεεν.

The original page is visually so dense and unvarying that users cannot identify which paragraphs contain reference information, and which contain task steps to help solve a problem. The page offers users no incentive to approach and read it, nor any clue to its organization or to the relative importance of the paragraphs. Before even attempting to rewrite the information, you can make it more approachable simply by placing it into structures that organize it into information types.

Revision

Ταβλε 1. Ινδιχατορσ οφ Στοραγε Σηορταγε Προβλεμσ

Ινδιχατορ	Προβλεμ	Μορε Ινφορματιον
Κ9Η	Ιφ ονε απλιχατιον ισ φλοοδινγ ανοτηερ ωιτη δατα ανδ τηε σεχονδ απλιχατιον ισ υναβλε το ρεχειϖε τηε δατα ανδ προχεσσ ιτ ατ αν αδεθυατε ρατε, τηε βυφφερσ βυιλδ υπ ιν ΜΧΑΤ στοραγε.	Τηισ σιτυατιον χαν βε αϖοιδεδ βψ υσινγ σεσσιον παχινγ.
ΙΣΤ999Ε μεσσαγε	ΜΧΑΤ ισ υναβλε το γετ ενουγη στοραγε το ισσυε α μεσσαγε	Αχχομπανιεσ τηε νορμαλ στοραγε σηορταγε μεσσαγε
ΦΓΗΜΠ=ξ϶0037϶ ωιτη ΘΦΡΔΤ=ξ϶0000϶	Α στοραγε σηορταγε ωηιλε ΜΧΑΤ ωασ ρεχειϖινγ δατα ορ σενδινγ α παχινγ ρεσπονσε.	Γιϖεν το ΔΔΦ ωηεν αν ΑΠΠΧΧΜΔ μαχρο ηασ χομπλετεδ
ΦΓΗΜΠ=ξ϶0093϶ ωιτη ΘΦΡΔΤ=ξ϶0000϶	Α στοραγε σηορταγε ωηιλε σενδινγ δατα. Υσυαλλψ μεανσ τηατ τηε σενδ ρεθυεστ ηασ τεμποραριλψ δεπλετεδ τηε βυφφερ ποολ το συχη αν εξτεντ τηατ τηε ποολ μυστ βε εξπανδεδ, ανδ τηε εξπανσιον ηασ νοτ οχχυρρεδ βεφορε τηε χομπλετιον οφ τηε ΑΠΠΧΧΜΔ μαχρο.	Γιϖεν το ΔΔΦ ωηεν αν ΑΠΠΧΧΜΔ μαχρο ηασ χομπλετεδ
ΣΝΑ σενσε χοδε ξ϶037Χ0000϶	Ινδιχατεσ α περμανεντ ινσυφφιχιεντ ρεσουρχε τηατ χουλδ βε στοραγε	
ΣΝΑ σενσε χοδε ξ϶300Α0000϶	Ινδιχατεσ α στοραγε τψπε προβλεμ, βυτ νοτ νεχεσσαριλψ α στοραγε σηορταγε.	
Π9Λ ΕΡΡΟΡ373 ορ ΕΡΡΟΡ30Α	Ινδιχατε στοραγε σηορταγε προβλεμσ βυτ αρε υνχομμον ιν ΜΧΑΤ.	

Πινποιντινγ τηε Σηορταγε

Ωηεν ανψ οφ τηεσε ινδιχατιονσ οφ α στοραγε προβλεμ ισ ρεχειϖεδ, τηε φολλοωινγ στεπσ χαν βε υσεδ το φινδ ουτ μορε ινφορματιον αβουτ τηε σηορταγε:

1. Δετερμινε τηε αρεα οφ στοραγε σηορταγε.

   Τηισ ισ ιμπορταντ, ασ τηε ΜΧΑΤ δισπλαψ χομ– μανδ ηασ ινφορματιον αβουτ τηε ΜΧΑΤ Κ9Η υσαγε. Τηισ δοεσ νοτ ηελπ διαγνοσισ ιφ τηε στοραγε σηορταγε ισ ιν ΜΧΑΤ πριϖατε. Ηοωεϖερ, Κ9Η υσαγε σηουλδ προβαβλψ βε χηεχκεδ ανψωαψ.

2. Δισπλαψ βυφφερ υσαγε.

   Ιφ τηε στοραγε σηορταγε ηασ οχχυρρεδ ιν ονε οφ τηε Κ9Η συβποολσ, τηεν α ∀Π ΣΕΤ,ΔΧςΡΦ∀ χομ–μανδ χαν βε υσεδ το δετερμινε τηε αμουντ οφ Κ9Η στοραγε βεινγ υσεδ βψ ΜΧΑΤ. Υσυαλλψ τηε βυφφερ υσε δισπλαψ γιϖεσ α γοοδ ιδεα οφ ωηιχη ΜΧΑΤ ποολ ισ χαυσινγ τηε στοραγε σηορταγε.

3. Μονιτορ βυφφερ υσαγε.

   Ωηεν λοοκινγ ατ α βυφφερ σηορταγε, ιτ ισ οφτεν ηελπφυλ το κνοω ωηετηερ τηε ονσετ οφ τηε προβ–λεμ ωασ γραδυαλ ορ ιμμεδιατε. Ιφ ρεγυλαρ βυφφερ υσαγε

In the revised page, the information has structure and organization. Even a cursory glance at the page now gives users a clear idea of how to approach the information. Headings and table captions, table rules, column gutters, and list numbers all help to guide users to and through the information, and white spaces give users places to pause. After pausing, users can more easily find where to resume reading.

Use visual cues to help users find what they need

You can help users find information quickly by using visual elements systematically and consistently.

Online interfaces often use icons to represent program and data objects. Well-designed icons can be easier to find than the names that they represent, and they can provide a lot of information about the characteristics of the objects that they represent and their relationships to other objects. You can also use iconic images on buttons to represent program tools or the actions of those tools.

In printed information, tabs or bleed bars (strips of color along the edge of a page) provide an easy way to help users find the beginning of a chapter or section. When users thumb through the book, the changing position of the tabs at the edge of the page tells them when they reach a new section. Running headings that include graphics along with the text of chapter headings and subheadings also help users to find information quickly and easily.

To help users to quickly find the information that they need, follow these guidelines:

- ❑ Visually identify recurring alternatives or contexts.
- ❑ Ensure that visual cues are usable in all environments.

Visually identify recurring alternatives or contexts

You can use icons in both online and printed information to show users what information is where. This technique can be particularly helpful when most of the information is common to several environments, scenarios, or user subgroups, but when some of the information applies to only a subset of the larger group.

Rather than repeat the common information for each subset, you can use a system of small graphics or icons, one for each unique situation, placed at the beginning of the subset information. This system helps users find the information that is relevant to them.

The following passage gives instructions for installing software from a CD-ROM on several operating systems:

Original

To install InfoManager:

1. Insert the CD in your computer's CD-ROM drive.
2. Follow the instructions for your operating system:
   a. To install on the Macintosh operating system:
      i. Open the CD-ROM drive.
      ii. Double-click **Install InfoManager now**.
   b. To install on the UNIX operating system:
      i. Log in as root user.
      ii. Mount the CD-ROM drive.
      iii. Change the directory to the mount point of your CD-ROM drive.
      iv. Type SETUPUNIX.
   c. To install on the Windows 2000 operating system:
      i. Click **Start**.
      ii. Select **Run**.
      iii. Type d:\SETUPWIN where d is the drive letter of your CD-ROM drive.
      iv. Click **OK**.
3. Follow the instructions in the setup program for your operating system.

In the original passage, users might need to read all of the text to find the information that applies to the operating system that interests them.

Revision

To install InfoManager:

1. Insert the CD in your computer's CD-ROM drive.
2. Follow the instructions for your operating system:

   Mac
   a. To install on the Macintosh operating system:
      i. Open the CD-ROM drive.
      ii. Double-click **Install InfoManager now**.

   UNIX
   b. To install on the UNIX operating system:
      i. Log in as root user.
      ii. Mount the CD-ROM drive.
      iii. Change the directory to the mount point of your CD-ROM drive.
      iv. Type SETUPUNIX.

   Windows
   c. To install on the Windows 2000 operating system:
      i. Click **Start**.
      ii. Select **Run**.
      iii. Type d:\SETUPWIN where d is the drive letter of your CD-ROM drive.
      iv. Click **OK**.
3. Follow the instructions in the setup program for your operating system.

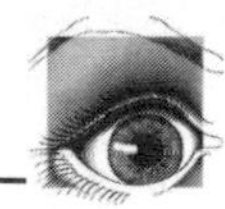

The revision adds operating system-specific icons to the left of the text, where users will see them first. Users can quickly find the information pertinent to their operating system, and then read those instructions, skipping the sections that apply to other operating systems. Thus they can move efficiently through the information without taking time to read the information that doesn't pertain to them.

Usually you want your users to read the text or instructions first, and then to look at illustrations to help them interpret the text or instructions. In such cases, place the graphic or illustration in one of the following arrangements:

- To the right of the associated text
- Below the text

In the following cases, place small graphics or illustrations before the text:

- The graphic helps users decide which text to read, as in the previous revised passage.
- Users need to understand the physical context in which they will find the objects or controls that you are about to describe.

Make the path obvious to your users. Ensure that they always know where to go next. Such visual guidance can be as simple as numbering the steps in a procedure or, in online information, highlighting links clearly and unambiguously.

Ensure that visual cues are usable in all environments

Users might view your information in contexts that you did not originally plan for. Ensure that the visual cues and retrievability aids that you add make your information accessible to as many of your users as possible. The following help topic includes links to other help topics, but the links are coded in color without underlining:

Original

Basic settings

You can customize the way that your text formatter behaves by changing the basic settings:

- Change the color scheme.
- Change the size of the Bookmark icon.
- Change the default fonts.
- Add words to your dictionary.

In black and white, the links disappear. If the window is displayed on a monochrome monitor or if the user is color-blind and cannot distinguish the color difference, the links might as well not exist.

Revision

Basic settings

You can customize the way that your text formatter behaves by changing the basic settings:

- ❑ Change the color scheme.
- ❑ Change the size of the Bookmark icon.
- ❑ Change the default fonts.
- ❑ Add words to your dictionary.

The revised example shows the same links with the color highlighting reinforced by underlining. The additional visual cue enables color-blind users and users of monochrome displays to distinguish such links from normal text.

You might be required to integrate your information into an environment where a different visual system is used. If your information is to be placed on the Web, for example, you might need to provide links to Web pages that represent links with different colors or highlighting. If you are faced with such a situation, choose the visual system that is either the most prevalent or the most obvious and usable. For example, the predominant highlighting for links on the Web is underscored blue text, so that style is an obvious choice.

Ensure that textual elements are legible

Be sure that textual information is always legible. The following factors are the major ones that can affect the legibility of text:

- Typeface
- Type size
- Contrast between text and background

You can ensure the legibility of both body text and text in graphics by following these guidelines:

- Use a legible typeface and size.
- Ensure that the text fits in the available space.
- Ensure that the contrast between text and background is adequate.

Use a legible typeface and size

Choose a text typeface and size that is appropriate for the particular combination of user, information type, and presentation medium to be used. A type size of 10 points is usually considered both economical and comfortable for adults to read in large blocks of print, as in a book. You can present some categories of information, such as figure and table captions, in smaller sizes if users don't need to read large blocks at a time.

For online information, text should appear the equivalent of 10 or 12 point, or no less than 1/8 inch high. Most online information delivery software is designed to display text at a readable size regardless of display resolution, but you should test your information at all supported display resolutions to verify its legibility. In any case, try not to use type smaller than 8 point. When space is not restricted, or when the primary audience is over 40 years old, use larger type for continuous text.

If users will view either printed or online information from farther away than about two feet, you must increase the type size to retain legibility. If your information is being presented in a multimedia demonstration at a trade show, for example, where users stand about six feet away from the display monitor, set your text to 32 points to ensure that it is legible.

Table 2 on page 312 shows the point sizes that are appropriate for various types of information:

Table 2. Minimum recommended type sizes.

Point size	Kind of information	How used
10	Concepts, task	Read from start to finish or at least in large, continuous blocks
9	Reference text	Read in limited blocks, usually a page or less, when looking for a small and specific subset of information
8	Tables, illustrations, sample code, syntax diagrams	In the context of a document element, read as small blocks of text: usually single sentences, short phrases, keywords, or lines of code
7 or 8	Footnotes, notes to tables	Read only once or infrequently
6 or 7*	Legal notices such as license agreements and copyright statements	Read only once or infrequently

* Law or your company's policies might require a specific type size.

In addition to choosing a legible type size, choose a simple, clean, well-proportioned typeface, especially for online information. The serifs on the main strokes of characters in many typefaces (such as the one that is used in the body of this book) help to lead the eye horizontally from one character to the next and thus facilitate the reading of continuous lines of text. The serifs on many typefaces, however, are very fine and can disappear from text that is displayed on most computer screens. For online information, therefore, select either of the following typefaces:

- A system standard, clean, sans-serif typeface such as Arial or Verdana
- A typeface with a strong, horizontal square or slab serif, such as Rockwell or Memphis

You can improve the effectiveness of illustrations by ensuring that the text in them is legible. In the following illustration, each object has a descriptive label:

Original

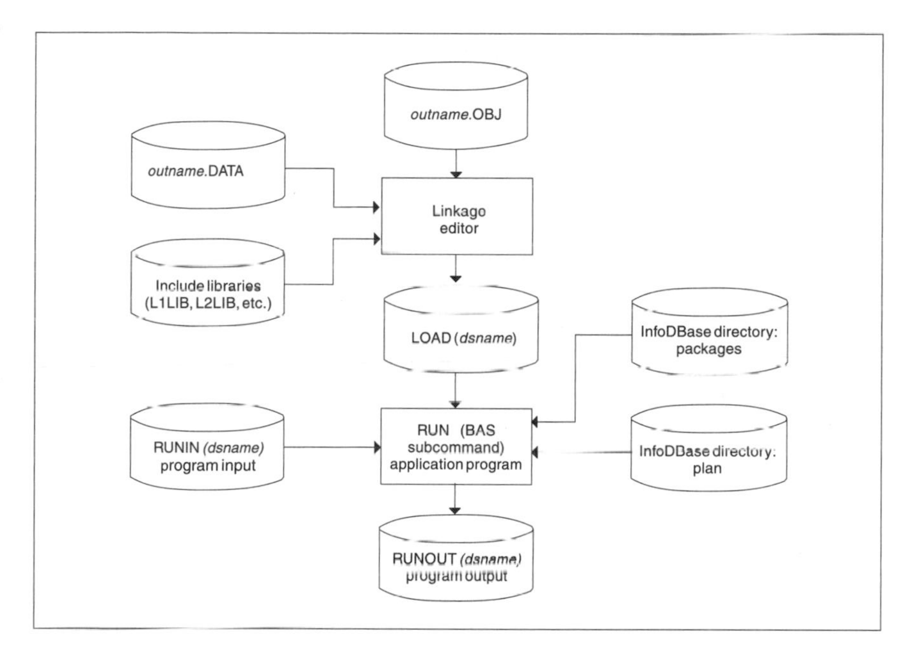

In the original illustration, the illustrator used 6 point text to fit a lot of information into a small space. As a result, many users will struggle to read it.

First revision

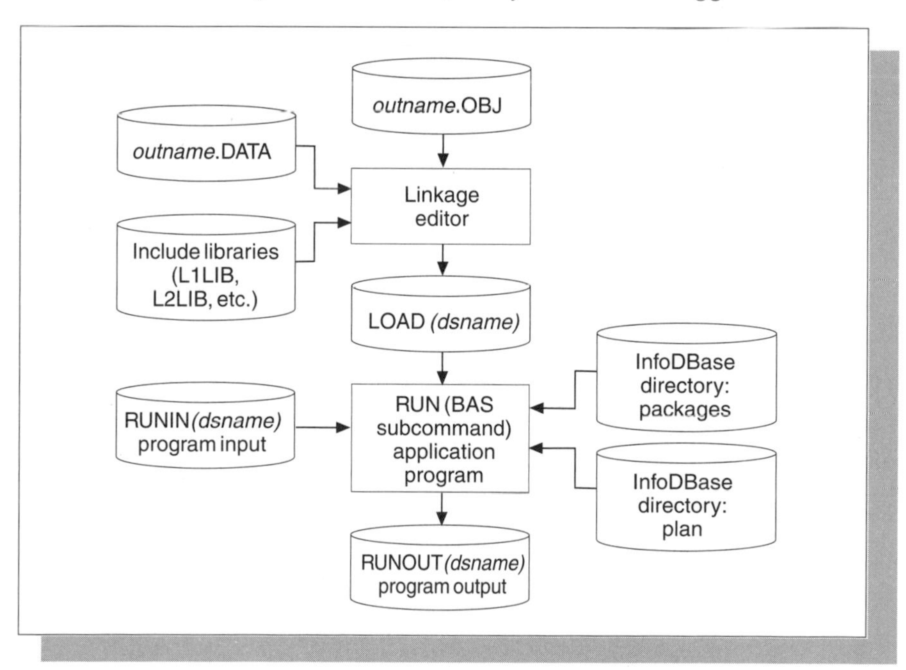

In the revised illustration, the illustrator made the text larger and therefore much more legible. The objects that contain the text are slightly larger, but the basic size of the illustration has not increased.

Here, the larger text comfortably fills many of the objects. When English is translated into other languages, however, the resulting text can occupy up to 30 percent more space. If your artwork will be translated, the illustrator must enlarge the objects accordingly to allow extra space for the additional text.

An alternative means of allowing for the expansion of translated text is to place all the text outside the objects so that it is not restricted by the objects' borders, as in the second revision:

Second revision

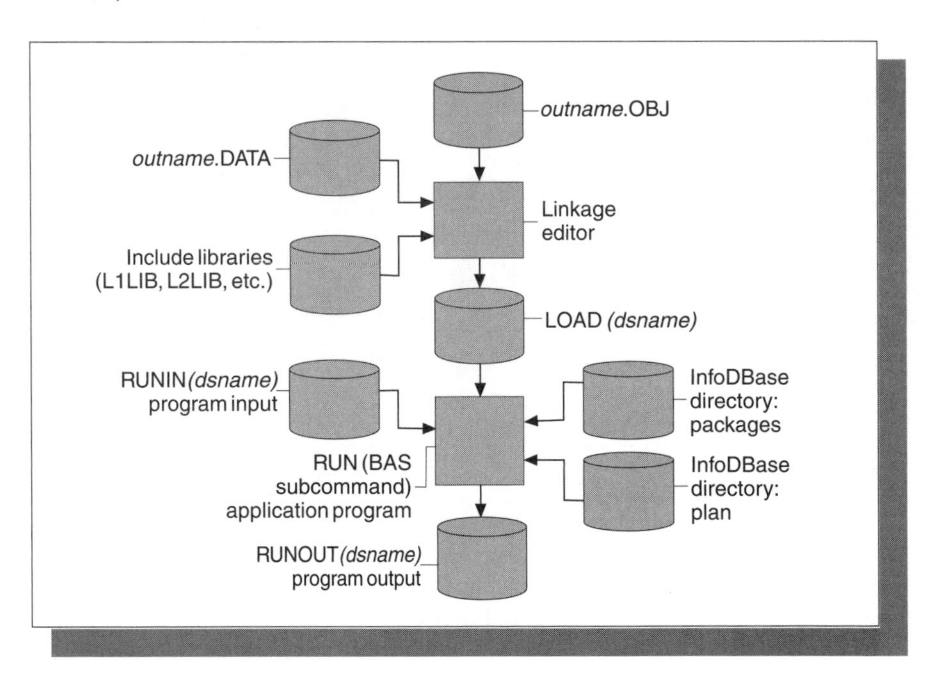

By placing the text outside the objects, as in the second revision, you can reduce the size of the objects to allow more space for text to expand into when it is translated. You can also add a fill color or shading to the objects without obscuring the text.

You can make your information easier to use by presenting it in a type size that is appropriate for the way it is used.

Ensure that the text fits in the available space

Small screens, such as those in pocket organizers, present a particular challenge in balancing legibility with space constraints. The following two English language screens from a pocket organizer pack a lot of information into a small space. Such an interface offers little room for essential information, even in English.

English version

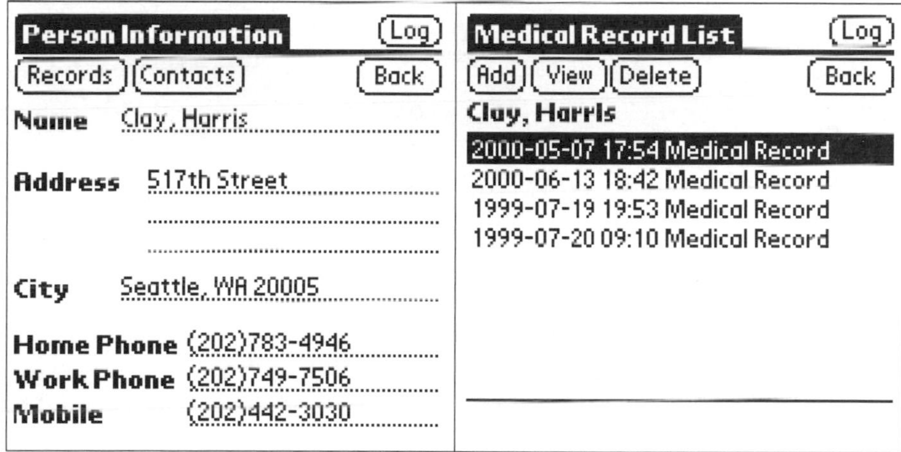

The English versions of the screens, though full, accommodate all of the elements comfortably.

Chinese version

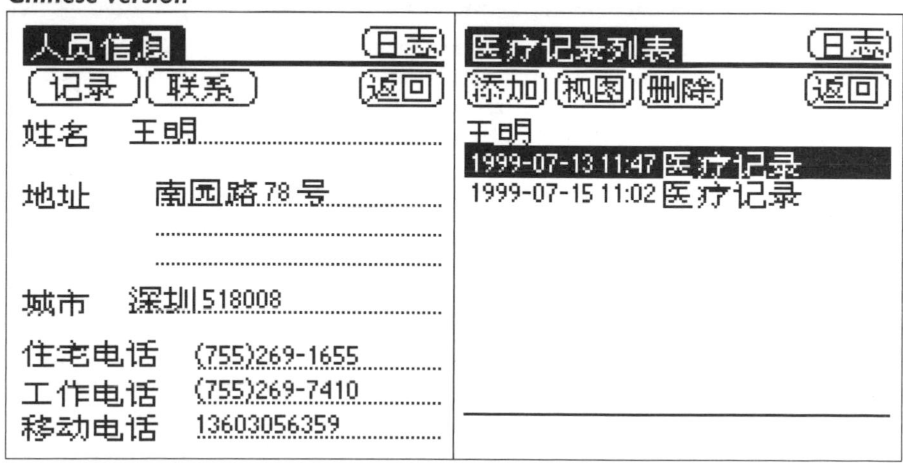

On equivalent screens in Chinese, because each character represents a word or word fragment, most of the text occupies less space than the English version.

German version

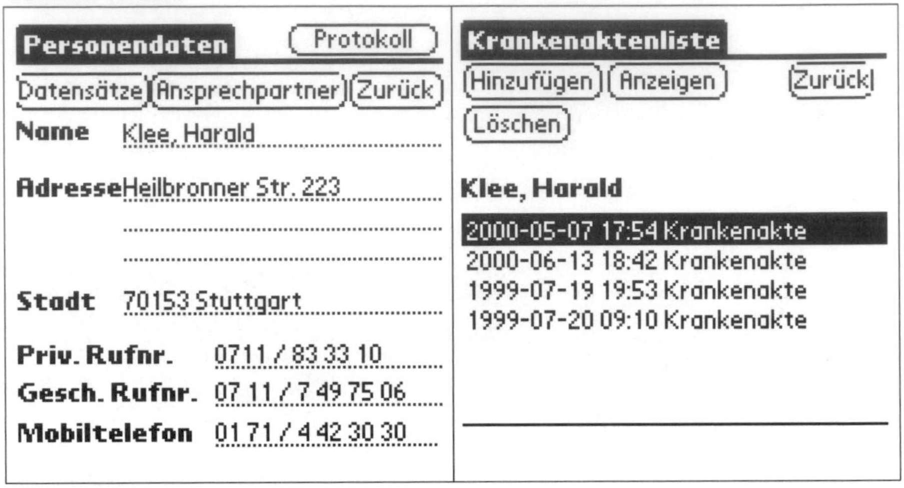

Personendaten Protokoll
Datensätze Ansprechpartner Zurück
Name Klee, Harald
Adresse Heilbronner Str. 223
Stadt 70153 Stuttgart
Priv. Rufnr. 0711 / 83 33 10
Gesch. Rufnr. 07 11 / 7 49 75 06
Mobiltelefon 01 71 / 4 42 30 30

Krankenaktenliste
Hinzufügen Anzeigen Zurück
Löschen
Klee, Harald
2000-05-07 17:54 Krankenakte
2000-06-13 18:42 Krankenakte
1999-07-19 19:53 Krankenakte
1999-07-20 09:10 Krankenakte

On the equivalent screens in German, however, where words are longer, text is crowded together, some labels are abbreviated, buttons are significantly larger, and, on the right, one of the buttons wraps to a second line.

To ensure that the design accommodates translation, find out ahead of time which languages the product will support, and try to design your visual elements so that they are large enough to accommodate languages that use more or longer words than English does to express the same information.

Ensure that the contrast between text and background is adequate

Another consideration for online text is the contrast between the text and the background. Ensure that, if you use colored text or colored background behind the text, contrast between them is strong enough that users can read the text. You might think that complementary colors, such as red and green, or orange and blue, provide the best contrast, but that is not necessarily true. In fact some color-blind people see red and green as the same color. You must ensure that the *values* of the colors, that is, the darkness or lightness of the colors, are different. You must test and verify that the colors you choose, if converted to greys, still contrast with one another well enough for the information to be clearly visible.

The following text box uses orange text on a blue background (complementary colors) to highlight an improvement in a database product:

Original

The shades of blue and orange, however, are so close to one another in value that, when they are converted to greys, the text almost disappears. Choosing a darker, more intense blue and a lighter (pale yellow) text color changes the contrast significantly.

Revision

With the revised color scheme, users have a much better chance of noticing the text box and reading and understanding its contents.

Use color and shading discreetly and appropriately

Color can be very helpful in establishing a hierarchy of importance, creating visual breaks, and clarifying information. Information that is to be presented online can usually include multiple colors with little impact to cost. In printed information, color can be expensive to use, so you might need to restrict its use. Even in black-and-white documents, though, shades of grey can be just as effective in adding clarity and interest to your information.

As with all graphic techniques, you should use color and shading to make the information easier to find, understand, and use. You must also ensure that the color or shading does not so overpower the information that it distracts or confuses the user.

Cool colors (blues and greens) recede, whereas warm colors (reds and oranges) advance. By placing cool and warm colors strategically, you can enhance the relative importance of information.

Never use color, though, as the only means of differentiating or highlighting information. You must ensure that the highlighting is still clear to users who cannot see the color, because of either altered color perception or blindness. About 10% of men of European origin have some form of color blindness. Some colors appear only as shades of grey to color-blind people, so your visual system should be easy to understand even when it is viewed in black and white. If you must use only one color (usually, but not always, black), you can use shading (tints, or screens, of black or the color) to help focus the user's attention on the most important information.

Consider, too, the possible international use of your information. The connotations of some colors differ from culture to culture. For example, red in a traffic sign is universally recognized to mean "stop" or "danger." A woman dressed in red, however, would be interpreted in some Asian cultures as a bride. If you use color to convey meaning, check with international experts to ensure that you don't use it in an ambiguous or potentially offensive way.

The following bar chart compares processor time to elapsed time for two functions of a database application:

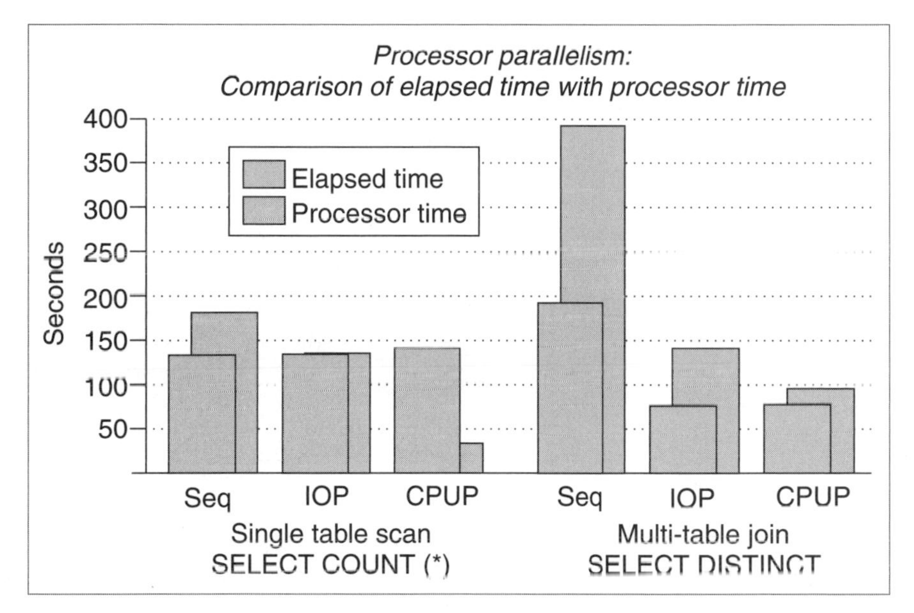

The original bar chart uses color to differentiate elapsed time from processor time. The colors that are used, however, are so close in value as to be indistinguishable when converted to shades of grey. Users cannot determine whether the bars for processor time are in front of or behind the bars for elapsed time. Thus, the original chart, with uniform shading, is almost meaningless.

First revision

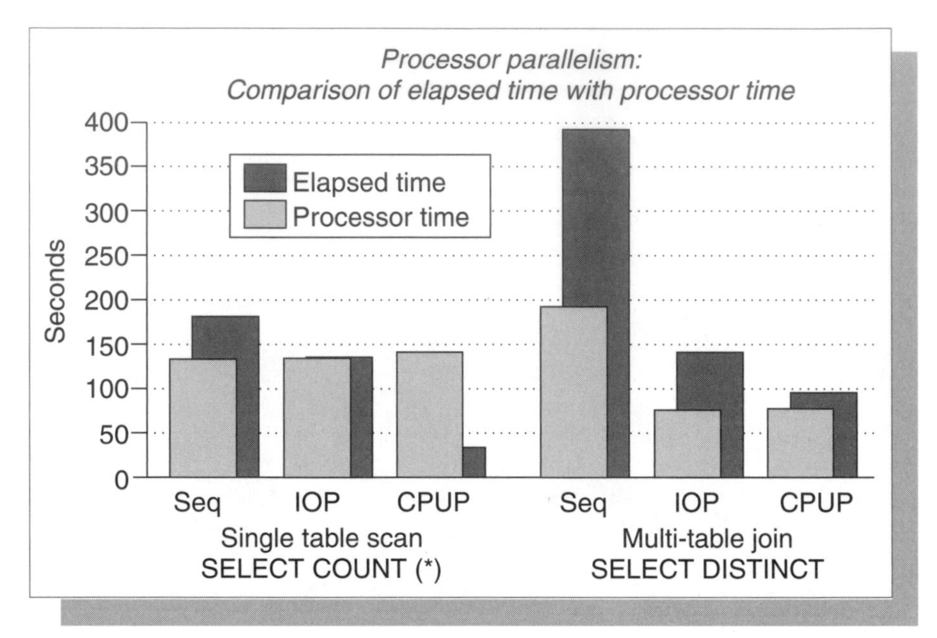

In the first revision, the colors have been changed so that their values differ significantly. Even when viewed in black and white, the bars for elapsed versus processor time are easy for users to differentiate.

In this version of the chart, the bars for elapsed time are much darker and more obvious than those for processor time, but they are placed behind the bars for processor time. Shading has been added to the revised chart to differentiate the bars for elapsed time from the bars for processor time. The placement and shading of the two sets of bars could have any of several implications:

- Elapsed time is more important because its bars are darker.
- Processor time is more important because its bars are in front.
- Neither statistic is more important than the other.

The combination of shading and bar position in the first revised chart does not clearly emphasize one set of statistics. If you do not want to emphasize one set of statistics more than the other, the design works as intended. If, however, you want to emphasize one set of statistics, a different design makes a stronger statement, as shown in the second and third revised charts:

Second revision

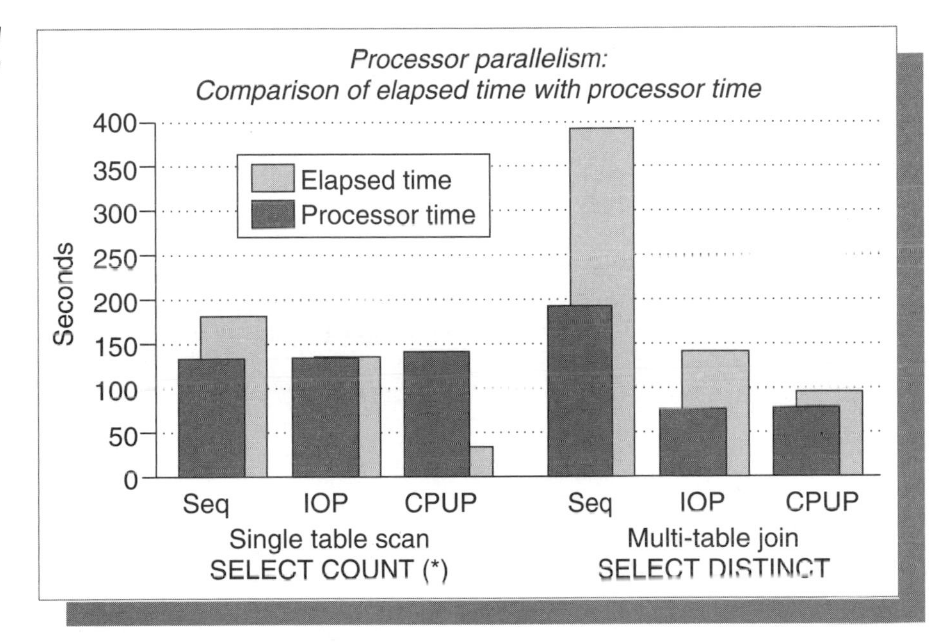

The darker, stronger shading in the second revised chart is on the set of bars in front; this design clearly emphasizes processor time. If you want to emphasize elapsed time, you need to revise how you use color (shading).

Third revision

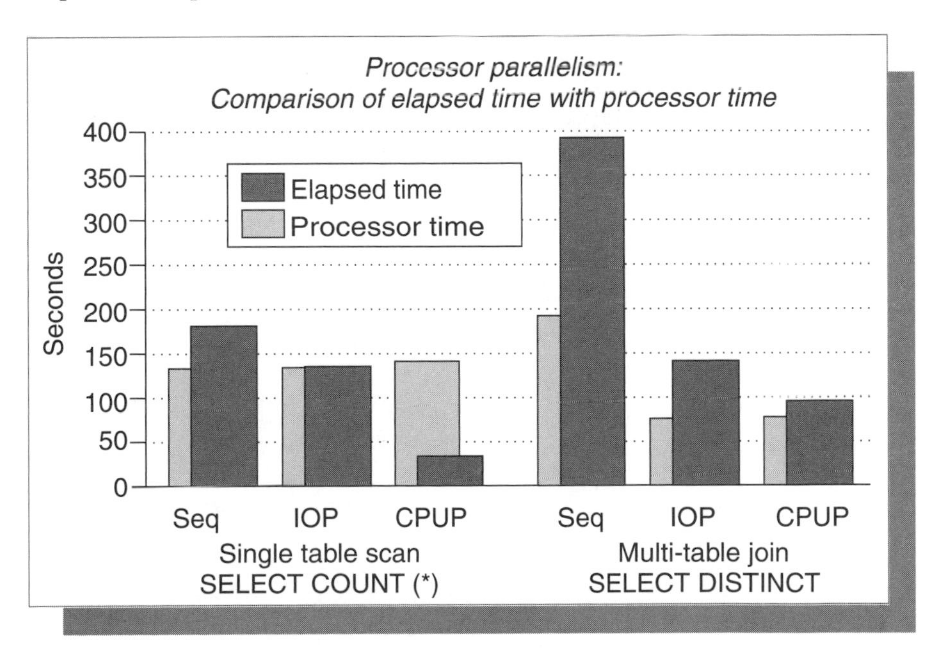

The third revised chart emphasizes elapsed time not only by using the darker shading for its bars, but also by moving its bars to the front. Color (shading) has been combined with position to strengthen the message.

When you select color and shading, consider the context within which your users will view your information. If, for example, your information will be made available on the Web and will have links to other information, use the same color coding for links that is used in associated Web documents, or provide an obvious key that explains the color scheme that you use.

Test your information and its colors in as many of the environments where users will view it as possible. Differences in Web browsers, pixel resolution, monitor specifications, color depth, room lighting, operating environment, and users' visual acuity can all affect how colors display and are perceived. You must ensure that the color scheme that you established has the effect that you intended.

Ensure that all users can access the information

The adage "a picture is worth a thousand words" does not take into account the many people who are visually impaired. For these users, a picture that is inadequately supported by text can leave a hole in their understanding of the information. Visual elements and techniques that enhance the information for a sighted user can actually impede a blind person's use of the same information.

An illustration or screen capture cannot be translated into Braille, so it is useless to a blind user. Therefore you must provide an explanation of the graphic, so that blind users can get the information through a screen reader.

The following illustration on a Web page represents a particular arrangement of XML data in a database. A user has moved the cursor over the image, causing the text on the ALT attribute of the hypertext IMG tag to appear.

Original

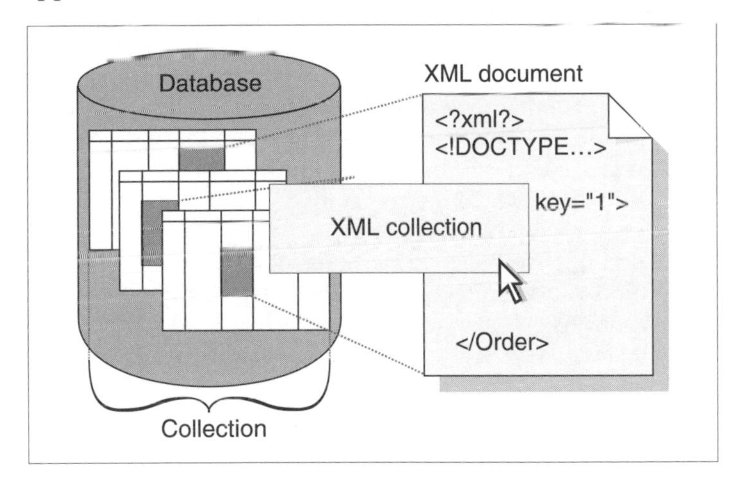

The original illustration shows alternate text that describes the graphic only very briefly. The description does not assure blind users that they have all of the important information.

Revision

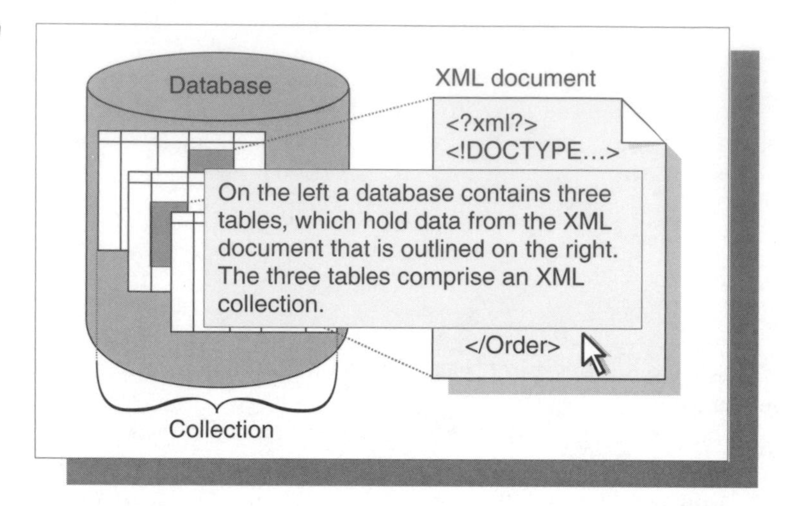

The alternate text in the revised illustration describes the illustration. From the description, visually impaired users can determine that the graphic illustrates a concept that is already described in the text.

Tables can also pose a problem for blind users. Some screen readers, which translate text into Braille or voice, scan across a line before going to the next. If text in a table cell wraps to a second line, the screen reader moves to the next cell before it reads the second line of the current cell.

You need not avoid using tables entirely, because screen readers and document composition software are constantly being improved to better handle such situations. Make your tables as simple as possible and avoid using complex layouts such as cells that span two or more rows or columns. If you are using a markup language such as HTML or SGML, you can use table row and cell coding techniques that help to define the table structure for visually impaired users.

If possible, avoid using tables to achieve specific layout effects. Screen readers tell their users the grammatical or document structure that is being used so that the users have an idea of how the information is organized. A table that does not present related, tabular information can be very confusing. For the same reason, do not replace such simple characters as bullets with graphics or icons unless you can do so without changing the markup of the list structure into a series of paragraphs.

In sum

Visual effectiveness can enhance all of the other quality characteristics. By itself, it can do little to improve the quality of your information, but if you use it as a vehicle to aid in improving each of the other quality characteristics, you can enliven your information and make it easier to use, understand, and find. Information that is visually effective helps the structure of the information disappear and thus focuses users' attention on the information itself.

Use the guidelines in this chapter to help ensure that your technical information is visually effective. The examples in the chapter represent only some of the possible practical applications of these guidelines.

When you review technical information for visual effectiveness, you can use the checklist on page 326 in two ways:

- ❑ As a reminder of what to look for, to ensure a thorough review
- ❑ As an evaluation tool, to determine the quality of the information

You can apply the quality rating in the third column of the checklist to the visual effectiveness guideline listed in the first column. Based on the number and severity of visual effectiveness problems you find, decide how the information rates on each guideline. You can then add your findings to "Quality checklist" on page 387, which covers all the quality characteristics.

Although the guidelines are intended to cover all areas for visual effectiveness, you might find additional items to add to the list for a guideline.

Guidelines for visual effectiveness	Items to look for	Quality rating
Use graphics that are meaningful and appropriate.	• Graphics make the text more meaningful. • Significant concepts are illustrated. • Graphics show only what users need to see.	1 2 3 4 5
Choose graphics that complement the text.	• Graphics provide significant information that users need. • Graphics help provide an overview of important concepts. • Graphics illustrate the information from the user's point of view.	1 2 3 4 5
Use visual elements for emphasis.	• Visual elements emphasize the appropriate information. • Visual elements do not distract.	1 2 3 4 5
Use visual elements logically and consistently.	• Heading hierarchy is visually logical and consistent. • Visual elements are presented and positioned consistently. • General look and feel is clean and consistent. • Symbols, icons, and navigational controls are used consistently.	1 2 3 4 5
Balance the number and placement of visual elements.	• Text is broken up by white space or varied structures. • White space appears intentional, not accidental.	1 2 3 4 5
Use visual cues to help users find what they need.	• Visual elements help guide users to correct paths. • Visual cues are unambiguous. • Visual cues are usable in all environments.	1 2 3 4 5
Ensure that textual elements are legible.	• Typeface and size is appropriate for the delivery medium and reading distance. • Text fits in the available space. • Sequential text is in a legible typeface and size. • The contrast between text and background is sufficient.	1 2 3 4 5

Guidelines for visual effectiveness	Items to look for	Quality rating
Use color and shading discreetly and appropriately.	• Color or shading emphasizes the appropriate information. • Colors provide sufficient contrast for color-blind users. • Color and shading are consistent. • Color does not convey meaning that might be ambiguous or offensive in another culture.	1 2 3 4 5
Ensure that all users can access the information.	• Graphics support, rather than replace, text. • Figure captions and alternate text adequately describe graphics. • Grammatical and document structures such as tables are simple and used appropriately.	1 2 3 4 5

Note: The scale for the quality rating goes from very satisfied (1) to very dissatisfied (5).

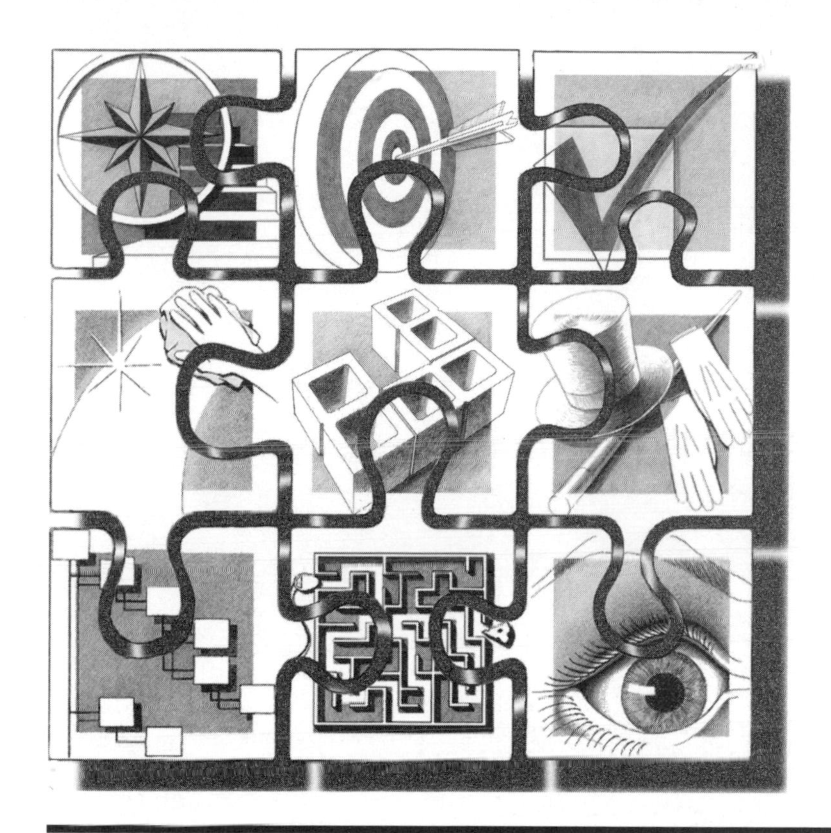

Part 4

Putting it all together

Developing quality technical information involves all of the quality characteristics. The chapters in this part of the book look at the interplay of characteristics. One of the chapters also looks at the roles of other people besides writers in the development cycle—technical reviewers, users, human factors engineers, testers, editors, and visual designers.

Chapter 12. Reviewing, testing, and evaluating technical information ... 357

Chapter 11

Applying more than one quality characteristic

The nine quality characteristics overlap. For instructional purposes, this book presents each quality characteristic in a separate chapter, but you will probably apply the guidelines for more than one quality characteristic to a piece of information. With practice, you can discern the effects of each quality characteristic and how to make improvements.

A good index, for example, must be accurate, complete, well organized, clear, and stylistically consistent. A good index contributes primarily to retrievability but also to completeness and task orientation (by including tasks as index entries). These are just some of the ways that the quality characteristics blend.

You can best improve a piece of writing by applying the guidelines for more than one quality characteristic. This chapter gives examples of the quality characteristics to apply in the following situations:

- Restructuring information into task, concept, and reference topics
- Writing for an international audience
- Writing for the Web

This chapter also provides guidance for revising technical information in general.

Applying quality characteristics to task information

Task information provides step-by-step instructions, along with information such as the rationale for the task, prerequisites, and examples. This type of information is common for software or for other products or situations where users need instructions.

Consider how you might improve the following task information:

Original

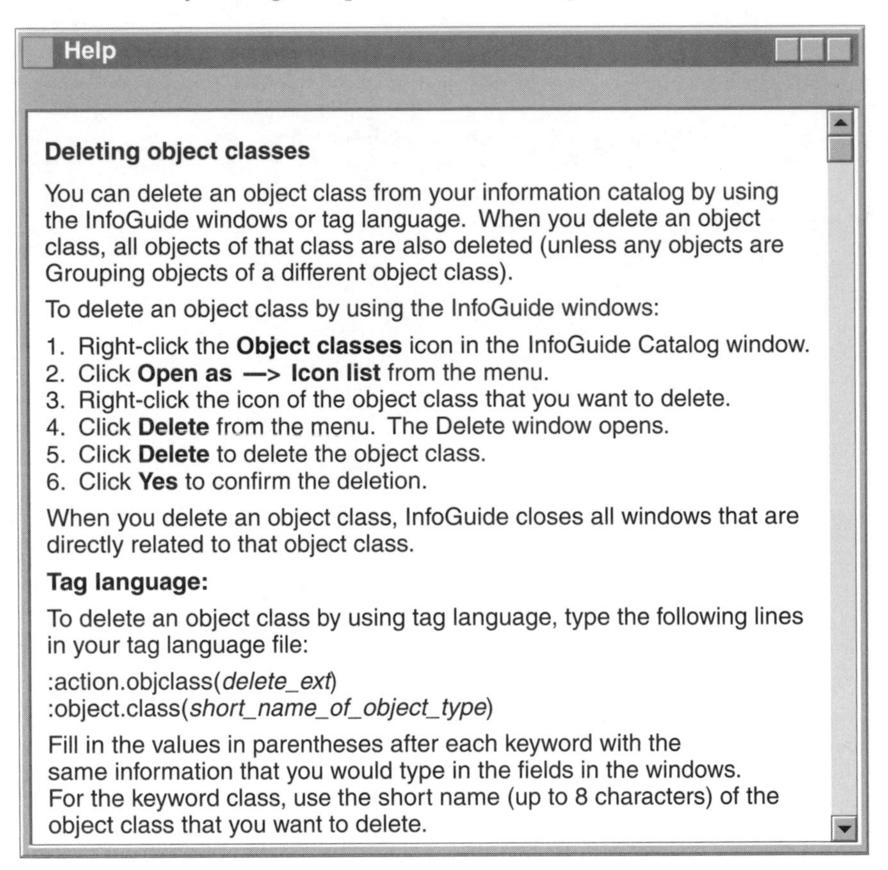

Help

Deleting object classes

You can delete an object class from your information catalog by using the InfoGuide windows or tag language. When you delete an object class, all objects of that class are also deleted (unless any objects are Grouping objects of a different object class).

To delete an object class by using the InfoGuide windows:

1. Right-click the **Object classes** icon in the InfoGuide Catalog window.
2. Click **Open as —> Icon list** from the menu.
3. Right-click the icon of the object class that you want to delete.
4. Click **Delete** from the menu. The Delete window opens.
5. Click **Delete** to delete the object class.
6. Click **Yes** to confirm the deletion.

When you delete an object class, InfoGuide closes all windows that are directly related to that object class.

Tag language:

To delete an object class by using tag language, type the following lines in your tag language file:

:action.objclass(*delete_ext*)
:object.class(*short_name_of_object_type*)

Fill in the values in parentheses after each keyword with the same information that you would type in the fields in the windows. For the keyword class, use the short name (up to 8 characters) of the object class that you want to delete.

This topic is too long. The problem is too much information, largely because of using one topic for both the primary audience for the information (users of the interface) and the secondary audience (users of the tag language).

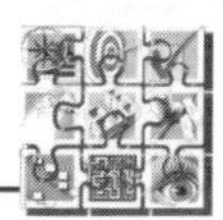

You might also think of this as an organization or retrievability problem: not breaking down the information into chunks that are small enough for easy scanning and reading. The original arrangement of the information also affects the task orientation by using the same topic to provide information for different audiences.

First revision

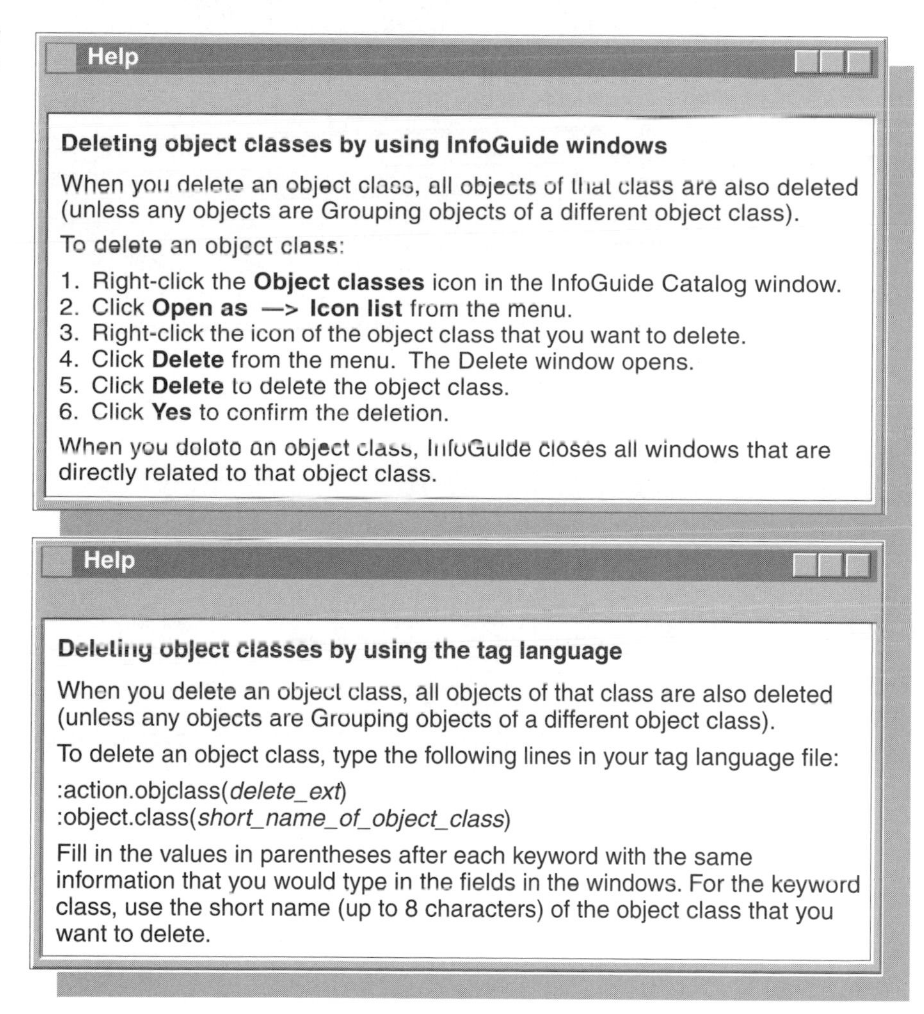

Help

Deleting object classes by using InfoGuide windows

When you delete an object class, all objects of that class are also deleted (unless any objects are Grouping objects of a different object class).

To delete an object class:

1. Right-click the **Object classes** icon in the InfoGuide Catalog window.
2. Click **Open as —> Icon list** from the menu.
3. Right-click the icon of the object class that you want to delete.
4. Click **Delete** from the menu. The Delete window opens.
5. Click **Delete** to delete the object class.
6. Click **Yes** to confirm the deletion.

When you delete an object class, InfoGuide closes all windows that are directly related to that object class.

Help

Deleting object classes by using the tag language

When you delete an object class, all objects of that class are also deleted (unless any objects are Grouping objects of a different object class).

To delete an object class, type the following lines in your tag language file:

:action.objclass(*delete_ext*)
:object.class(*short_name_of_object_class*)

Fill in the values in parentheses after each keyword with the same information that you would type in the fields in the windows. For the keyword class, use the short name (up to 8 characters) of the object class that you want to delete.

The revised information is split into two topics, one for the interface windows and one for the tag language. Each set of users can choose which topic to go to, whether they are linking from another topic, from the results of a search, from the index, or from the table of contents. Neither set of users needs to read or scroll past information that does not interest them.

Now you can refine each topic, eliminating internal redundancies or ambiguities. You might need to delete some information and to add or change some other information.

Second revision

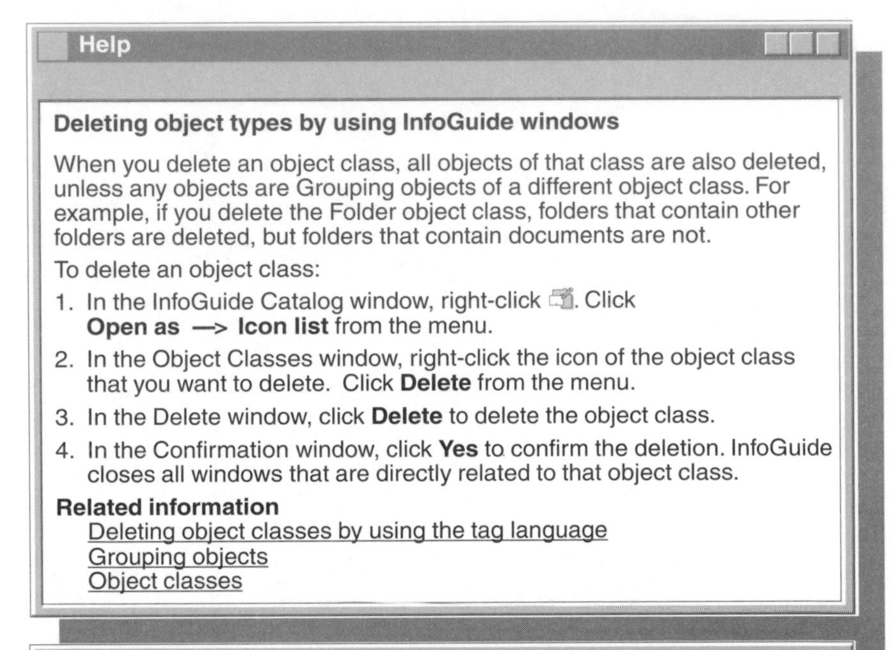

Help

Deleting object types by using InfoGuide windows

When you delete an object class, all objects of that class are also deleted, unless any objects are Grouping objects of a different object class. For example, if you delete the Folder object class, folders that contain other folders are deleted, but folders that contain documents are not.

To delete an object class:

1. In the InfoGuide Catalog window, right-click [icon]. Click **Open as —> Icon list** from the menu.
2. In the Object Classes window, right-click the icon of the object class that you want to delete. Click **Delete** from the menu.
3. In the Delete window, click **Delete** to delete the object class.
4. In the Confirmation window, click **Yes** to confirm the deletion. InfoGuide closes all windows that are directly related to that object class.

Related information

Deleting object classes by using the tag language
Grouping objects
Object classes

Help

Deleting object classes by using the tag language

When you delete an object class, all objects of that class are also deleted, unless any objects are Grouping objects of a different object class. For example, if you delete the Folder object class, folders that contain other folders are deleted, but folders that contain documents are not.

To delete an object class:

1. Determine the short name of the object class that you want to delete. You can find the short name in the description of the object class.
2. Enter the following lines in your tag language file, using the appropriate short name:

   :action.objclass(*delete_ext*)
   :object.class(*short_name_of_object_class*)
3. Run the file.

Related information

Deleting object classes by using the InfoGuide window
Grouping objects
Object classes
Tag-language syntax

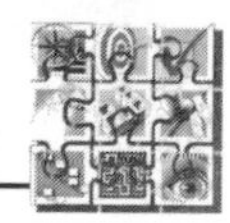

The second revision addresses problems that are related to the following quality characteristics:

- Task orientation, accuracy, and completeness

  The second revision replaces the reference information with steps that fit the task.

  The original information for using the tag language mentions filling in the values in parentheses as a user would type the values in fields in the interface. However, the task information for using the interface indicates that users need to select items, not type something. The second revision eliminates this inaccurate reference.

  The second revision consistently states the location for the action that the user is to do in the interface (using a pattern for what to present in the topic). This information could instead be stated as the result of each action (such as "The Delete window opens"), but this addition might annoy experts. Keep in mind the level of expertise of your audience and whether such changes are appropriate.

- Clarity, style, and concreteness

  The significance of Grouping objects was unclear in the original help topic and in the first revision. The second revision links to a definition of the term and also to related information about the concept of object classes.

  The second revision includes the style change of combining the click step and the select step into one step, to encourage thinking of these as one action.

  The second revision adds an example to help explain how the Grouping object behaves.

- Organization, retrievability, and visual effectiveness

  Breaking one long topic into two helps to make each set of information more effective for each set of users.

  The links to related information ensure that users are not at a dead end when they arrive at this topic. If they need more information about a concept or about the tag language syntax, they can easily link to it. Users who are interested in learning the other way of deleting object classes or who somehow landed at the wrong task can link to the other task.

  The second revision makes clear what the steps are in the task. It also uses a picture of the icon rather than the verbal description of it; alternative text for the image would ensure that screen readers could provide vision-impaired users with the meaning.

The following guidelines apply to these revisions of task information:

Quality characteristic	Relevant guidelines for task information
Task orientation	Write for the intended audience. Use headings that reveal the tasks. Provide clear, step-by-step instructions.
Accuracy	Maintain consistency in all information about a subject.
Completeness	Use patterns of information to ensure proper coverage.
Clarity	Define each term that is new to the intended audience.
Concreteness	Use focused, realistic, accurate, up-to-date examples.
Style	Create and follow style guidelines.
Organization	Organize topics for quick retrieval. Organize information consistently.
Retrievability	Design helpful links.
Visual effectiveness	Use visual cues to help users find what they need. Ensure that all users can access the information.

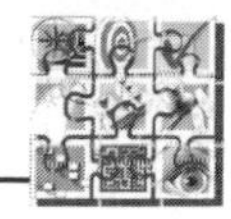

Applying quality characteristics to conceptual information

Conceptual information provides the background that users (especially novices) need before they can understand something or successfully work with it. Concepts can help users feel comfortable with new information and relate it to what they already know. The following subjects, for example, could be treated as concepts:

- The process of how something works
- An overview of a product, parts of a product, or several products
- A comparison of related technologies
- An abstraction that supports a function

The following passage is about the process of managing the presentation of help topics in response to actions by users. This description is of interest to developers of the software for a help engine or to the people who provide service for this software.

Original

> A link is established for each function request to the help engine. Each link has its own "current" help. Help selection from one link does not affect the current help on another link. Any link may select any help; more than one link may select the same help.
>
> The default help, 0, is considered current for the link unless changed by request on that link. The default is used before any help is created or selected on the link, or if the link explicitly deletes its own current help.

This passage is hard to understand. The initial hurdle seems to be the passive voice, which masks the agent. Who does what? When do they do it? The following revision deals with these problems:

First revision

> The help engine creates a link for each request that requires processing. Each link connects to a help topic. Connecting to a help topic does not change the connection between any other link and help topic. Any link can connect to any help topic; more than one link can connect to the same help topic.
>
> Initially, a link connects to the default help topic. The help engine can connect the link to a new help topic, or delete the help topic that the link connects to. If the help engine deletes the help topic, the engine also resets the link to connect to the default help topic.

The first revision uses an active style and clarifies when the help processor is responsible for an action. Each sentence is easier to understand, but the meaning of the passage is still hard to find. The passage is still basically two dull paragraphs. To make the meaning obvious, you can apply some guidelines from retrievability and visual effectiveness, as in the second revision:

Second revision

Process of linking help topics

In response to actions by users who request the help, the help engine takes the following actions:

Action by user	Action by help engine
No action	Connects to the default help topic
Requests help	Creates a link to the appropriate help topic
Requests different help	Links to a new help topic
Leaves help	Returns to the default help topic

The following principles apply to links between help topics:

- Any link can connect to any help topic.
- More than one link can connect to the same help topic.
- Connecting a link to a help topic does not change the connection between any other link and help topic.

The second revision uses a heading, a table, and a list to show at a glance what the information is and how the parts relate to each other. The passage now can stand as a discrete topic. You might decide that the topic needs to define some terms (such as *link* and *default help topic*) or link to a concept topic that explains these terms. Examples would also help make the passage more concrete and therefore easier to understand.

The following guidelines apply to these and potential revisions:

Quality characteristic	Relevant guidelines for conceptual information
Task orientation	Present information from the user's point of view.
Completeness	Use patterns of information to ensure proper coverage.
Clarity	Present similar information in a similar way. • Use lists appropriately. • Segment information into tables. Define each term that is new to the intended audience.
Concreteness	Choose examples that are appropriate for the audience and subject. Use general language appropriately.
Organization	Emphasize main points; subordinate secondary points.
Retrievability	Design helpful links.
Visual effectiveness	Use visual cues to help users find what they need.

Applying quality characteristics to reference information

Reference information provides quick access to facts, often as lists or tables. The following passage uses a list as the main organizer:

Original

> The keywords are:
>
> - ❑ The component identification keyword
>
>   This is the first keyword in the string. A search of the database with this keyword alone would detect all reported problems for the product.
>
> - ❑ The type-of-failure keyword. The second keyword specifies the type of failure that occurred. Its values can be:
>
>   — ABENDxxx
>   — ABENDUxxx
>   — DOC
>   — INCORROUT
>   — MSGx
>   — PERFM
>   — WAIT/LOOP
>
> - ❑ Symptom keywords
>
>   These can follow the keywords above and supply additional details about the failure. You select these keywords as you proceed through the type-of-failure keyword procedure that applies to your problem.
>
>   The suggested approach is to add symptom keywords to the search argument gradually so that you receive all problem descriptions that might match your problem. You can AND or OR additional keywords in various combinations to the keyword string to reduce the number of hits.
>
> - ❑ Dependency keywords
>
>   These are program- or device-dependent keywords that define the specific environment in which the problem occurred. When added to your set of keywords, they can help reduce the number of problem descriptions that you need to examine. See Appendix B, "Dependency Keywords," on page 303 for a list.

The original passage highlights the keywords, but the rest of the information is hard to sort out. The brief descriptions of the keywords leave you wondering "So what?" Buried in some of the paragraphs is information about how to use a keyword. However, you must read carefully to find this information and then reorient it to something for a user to do. The passage also appears to be part of a larger whole because it has no title.

You might decide that this passage communicates reference information about keyword strings and can stand on its own as a topic. You can there-

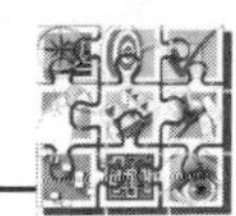

fore give it a noun as an appropriate title, add a definition, and organize the information to clarify the types of keywords and their general use. You can also link to the related reference information about dependency keywords.

First revision

Keyword strings

A *keyword string* is a set of words that you use to describe a problem with the product when you report a problem to software support.

Type of keyword	Description or value	Use this type of keyword to . . .
Component identification	A set of characters that represents the product or an orderable feature of the product	Find all reported problems with the product or with an orderable feature
Type of failure	❑ ABENDxxx ❑ ABENDUxxx ❑ DOC ❑ INCORROUT ❑ MSGx ❑ PERFM ❑ WAIT/LOOP	Refine your search to a particular type of failure for the product or orderable feature
Symptom	Details about the failure	Refine your search gradually (by combining the symptom keywords in various ways) so that you receive all problem descriptions that might match your problem
Dependency	Program- or device-dependent keywords that define the environment in which the problem occurred	Help reduce the number of problem descriptions that you need to examine

Related reference

Dependency keywords

The first revision makes the meaning obvious by its structure. The revision also presents the information from the user's point of view rather than from the point of view of the product. The information in the revision not only looks easier to understand; it *is* easier to understand. While turning the passage into a discrete topic, you have improved the clarity, task orientation, retrievability, and visual effectiveness of the original passage.

You have also improved the completeness of the original passage by filling in gaps that the pattern of information revealed. The original passage did not describe the first keyword, other than to say that it comes first. The table now conveys this information about sequence implicitly, through the order of the keywords from top to bottom. The original passage also did not give a use for the type-of-failure keyword. The table in the first revision includes both of these items. The following revision includes another improvement:

Second revision

Keyword strings

A *keyword string* is a set of words that you use to describe a problem with the product when you report a problem to software support.

Type of keyword	Description or value	Use this type of keyword to . . .	Example of a keyword string
Component identification	A set of characters that represents the product or an orderable feature of the product	Find all reported problems with the product or with an orderable feature	CKJPROD
Type of failure	❑ ABENDxxx ❑ ABENDUxxx ❑ DOC ❑ PERFM ❑ MSGx ❑ INCORROUT ❑ WAIT/LOOP	Refine your search to a particular type of failure for the product or orderable feature	CKJPROD WAIT
Symptom	Details about the failure	Refine your search gradually (combining the symptom keywords in various ways) so that you receive all problem descriptions that might match your problem	CKJPROD WAIT CYCLE
Dependency	Program- or device-dependent keywords that define the environment in which the problem occurred	Help reduce the number of problem descriptions that you need to examine	CKJPROD WAIT CYCLE AS4

Related reference

Dependency keywords

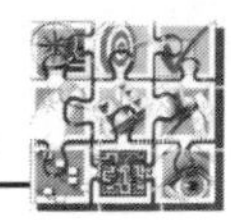

The second revision adds a column for examples. These examples help users see what the rest of the information is about and relate the information to a keyword string of their own. You have made the topic more concrete by adding the examples.

The revision reflects the following characteristics and guidelines:

Quality characteristic	Relevant guidelines for reference information
Task orientation	Present information from the user's point of view. Indicate a practical reason for information.
Completeness	Use patterns of information to ensure proper coverage.
Clarity	Present similar information in a similar way. • Use lists appropriately. • Segment information into tables. Define each term that is new to the intended audience.
Concreteness	Choose examples that are appropriate for the audience and subject. Make examples easy to find.
Organization	Organize information into discrete topics by type. Emphasize main points; subordinate secondary points.
Retrievability	Design helpful links.
Visual effectiveness	Use visual cues to help users find what they need.

Applying quality characteristics to information for an international audience

Whether your audience is in North America or also on other continents, most guidelines for the quality characteristics apply equally well. If your information is on the Web, users anywhere might have access to it.

To make your information easy to translate and easy to apply to other cultures, focus mainly on the quality characteristics that contribute to ease of understanding:

- Clarity
- Concreteness
- Style

In addition, some aspects of visual effectiveness require sensitivity to cultural differences.

If information is hard to understand, a translator probably cannot make it easy to understand. Very few translators have ready access to development groups or the time to get answers to the questions that the information might raise. If the translators perceive an ambiguity, for example, they might preserve it or select the meaning that they consider more likely, which might not be the right choice.

To reduce the time and cost of translation, many companies that require a significant amount of translation rely on a combination of human and machine translation. In such situations, the clarity of the information is even more important because the more costly human translators must do the work when machine translation tools cannot. In addition, reducing the amount of information has the side benefit of fewer words to translate.

Suppose that the marketing plan for a product that you work on is to sell the first release in the continental United States and to sell subsequent releases in more and more places throughout the world. This information will also be available on the Web. What preparations would you make so that you could

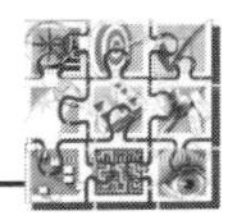

easily adapt the information to an international audience? Suppose that the following task information is being developed for such a product, with much input from marketing and development:

Original

> Congratulations on buying your global positioning system (GPS) device, InfoGPS! With InfoGPS you can always make a determination about where you are, and you can discover and locate the best route to wherever you want to go.
>
> InfoGPS was programmed at the factory with the date and time and the satellites that are appropriate there. However, chances are you're not at our factory! Therefore, you may need to reset the device for where you are, though the date is probably not incorrect in the continental United States. So first check the time to make sure it is correct, given the different time zones in the United States. InfoGPS does the rest for you. About five minutes are spent looking for the satellites it expects to find. If they are not found, the AutoInitialize search facility is started, which (since the latest satellite search technology was used) will take a maximum of five minutes to search for whatever satellites are available where you are. The next time you turn on InfoGPS, it will find its position in less than a minute unless you have moved more than 100 miles from the last location where it was used. For example, if you go from Chicago to Cincinnati, you must initialize InfoGPS again, but if you go from Chicago to Kenosha, you don't.
>
> Rather than wait for the AutoInitialize search facility, you can try initializing InfoGPS yourself. To initialize the device, press the Menu key. Then press the down arrow key until you see **Initialize GPS** highlighted. Then press the right arrow key to display the map screen. Use the arrow keys on the map screen to move the crosshairs to your approximate location. There are In and Out keys that you can use to resize the map and so find your location more easily. You can check the latitude and longitude of the cursor position in the box at the top. Once you are satisfied that the position of the crosshairs matches your position, press the Enter key.

The quality characteristics that primarily affect ease of use (task orientation, accuracy, and completeness) apply to technical information throughout the world. Regardless of where the InfoGPS product will be sold, the tasks and the steps must be clear. This passage looks as though it has no steps, because the steps are run into a paragraph. In addition, extraneous pieces of information (such as mentions of the factory, time zones, and search technology) mask the choices that a user has. The following guidelines for task orientation and completeness apply here:

- Present information from the user's point of view.
- Provide clear step-by-step instructions.
- Cover each topic in just as much detail as users need.

The tone of the passage has a marketing flavor. Marketing language tends to be idiomatic and therefore hard to translate and hard for nonnative speakers to understand. The passage has many other clarity and style problems that could confuse translators and nonnative speakers, such as:

- ❑ Roundabout expressions such as *make a determination about*
- ❑ Repetitive expressions such as *discover* and *locate* (which are also not as direct as *find*)
- ❑ Negative expression (*not incorrect)* instead of a positive expression
- ❑ Vague referent *they* instead of *these satellites*
- ❑ Missing *that* and *that are* (making the text harder to parse and translate)
- ❑ Ambiguous use of *since* instead of *because* and *once* instead of *after*
- ❑ Long noun strings (such as *satellite search technology* and *AutoInitialize search facility*)
- ❑ Use of *may* instead of *can* or *might*
- ❑ Passive voice where active voice would be clearer
- ❑ Misplaced modifier (*in the box at the top* should not modify *cursor position*)

In addition, these paragraphs do not provide visual cues such as lists that could help users skim the information.

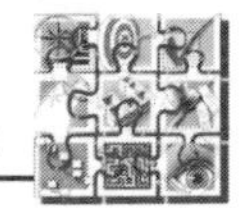

The following revision remedies many of these problems:

First revision

> When you first turn on InfoGPS, your global positioning system (GPS) device, it knows the current date but maybe not the local time and your position. Reset the time if it is incorrect. The satellites that InfoGPS automatically looks for are not necessarily the ones that are available where you are. If InfoGPS doesn't find these satellites after five minutes of searching, it starts looking for any satellites. This AutoInitialization takes a maximum of five minutes until InfoGPS finds the local satellites and shows your position. The next time that you turn on InfoGPS, it will find your position in less than a minute unless you have moved more than 100 miles from the last location where you used it. For example, if you go from Chicago to Cincinnati, you must initialize InfoGPS again, but if you go from Chicago to Kenosha, you don't.
>
> Rather than wait for the AutoInitialization, you can initialize InfoGPS yourself:
>
> 1. Press the Menu key.
> 2. Press the down arrow key until you see **Initialize GPS** highlighted.
> 3. Press the right arrow key to display the map screen.
> 4. Use the arrow keys to move the crosshairs on the map screen to your approximate location.
>
>   **Tip:** You can use the In and Out keys to resize the map and find your location more easily. In the box at the top, you can check the latitude and longitude of the cursor position.
> 5. When you are satisfied that the position of the crosshairs matches your position, press the Enter key.

The first revision fixed many of the task-orientation problems in the second paragraph but not in the first paragraph. The tone of the first paragraph is more appropriate than in the original passage, and many of the clarity and style problems have been fixed throughout, but the first paragraph is still hard to understand in terms of what to do and when. Imperative statements are mixed with descriptive statements.

If you analyze the task of establishing one's position, you find that a user must do this task more than once. The original passage suggested that the user could do this task at the outset and then the device would be set forever, but the information about going farther than 100 miles implies that the user would need to redo this positioning. Therefore, the task must be presented in a way that suits not only first-time users, but also users who have gone beyond the 100-mile radius where the old positioning applied. In fact, the latter situation is probably more common for users who need this information.

Terms that are used in a product interface can cause problems in using the product and also problems in understanding the information when it uses these terms. The name of the AutoInitialize feature might be difficult for non-native speakers and for translators to understand if they do not recognize *Auto* as an abbreviation for *automatic*. In addition, some users might not understand the technical term *Initialize*.

You need to work with developers and with human factors engineers to make the terminology in the interface acceptable. Such refining is important even in the first release of a product because a change in the interface later could confuse current users. The date format, for example, should be in the international format yyyy-mm-dd and should allow users to change it to their preferred format. You might also consider whether to start out using *kilometers* rather than *miles* as the distance metric.

Another issue to consider for the time when this product is sold throughout the world concerns the date. The date that is set at the factory could get out of sync with the local calendar. The instructions could start with a recommendation that users check both the date and the time.

The example is specific to the United States, but it seems unnecessary because the distance information is already specific. Besides, only people who are familiar with the area of the Midwestern United States that the example highlights might find the example helpful. Deleting the example is probably more appropriate than adapting it to every locale.

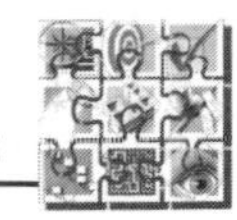

Second revision

Orienting your InfoGPS

When you turn on InfoGPS, your global positioning system (GPS) device, check the date and time to make sure that they are correct.

The satellites that InfoGPS is programmed to look for might not be the ones that are available where you are. You can choose either of two methods to acquire signals from the local satellites:

- Let InfoGPS search automatically.
- Set up your position yourself.

Regardless of the method that you use to orient your InfoGPS, the next time that you turn it on, it will find your position in less than a minute unless you have moved more than 100 miles from the last location where you used it.

Automatic searching for satellites

When you let InfoGPS manage its own searching for local satellites, it spends a maximum of 10 minutes:

- Up to 5 minutes to check for the satellites that it expects to find in the area that it was programmed for
- Up to 5 minutes to look for any local satellites

To activate automatic searching, simply turn on InfoGPS.

Manual searching for satellites

Rather than wait for automatic searching, you can orient InfoGPS yourself:

1. Press the Menu key.
2. Press the down arrow key until you see **Position GPS** highlighted.
3. Press the right arrow key to display the map screen.
4. Use the arrow keys to move the crosshairs on the map screen to your approximate location.

   Tip: You can use the In and Out keys to resize the map and find your location more easily. In the box at the top, you can check the latitude and longitude of the cursor position and the distance from the last known position.
5. When you are satisfied that the position of the crosshairs matches your position, press the Enter key.

The second revision clarifies the relationship between the automatic searching and the manual positioning. The tasks can work anywhere in the world. This revision also replaces *initialize* in the interface.

Apart from the task orientation and completeness guidelines that affect the revisions of this passage, the following guidelines are most relevant to improving this information for nonnative users of English and for translators:

Quality characteristic	Relevant guidelines for writing for an international audience
Clarity	Avoid ambiguity. Keep elements short. Use technical terms only if they are necessary and appropriate.
Concreteness	Choose examples that are appropriate for the audience and subject.
Style	Write with the appropriate tone. Use an active style.

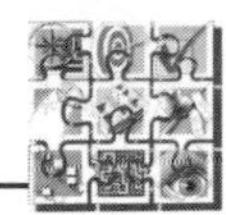

Applying quality characteristics to information on the Web

Almost any form of technical information can be distributed and accessed on the Web. Information that you write might be, for example, part of an information center or a product support site. Many software products use a Web browser as their interface.

Writing for the Web includes all of the considerations of writing for an international audience if people around the world might access your information. Such information might be translated dynamically for people who want the information in their own language and who have the capability to request that translation online.

In addition to ease of understanding, the ease of finding information is especially important on the Web. Users need to find the particular information that they need within the large amount of information that is usually available at a Web site. They also need to know how the pieces of information relate to each other and perhaps to information at other sites.

Consider how you might organize the set of information for InfoGPS, the product with the task topic in "Applying quality characteristics to information for an international audience" on page 344. Suppose that all of the task, concept, and reference topics for InfoGPS have been identified as follows:

Task topics	Concept topics	Reference topics
Installing InfoGPS	Global positioning system (GPS)	Restrictions
Turning on InfoGPS	Components of a route	Keys
Setting the date and time		
Setting options		
Orienting your InfoGPS		
Creating a route		
Following a route		
Changing a route		
Deleting a route		

The task of orienting the InfoGPS is not an isolated task. Users might have questions about related tasks such as setting the date and time. Similarly, they might want to learn what a global positioning system is (a concept topic) or what a particular key does (a reference topic). You can therefore add the following links at the bottom of the task:

Related tasks
Turning on InfoGPS
Setting the date and time

Related concepts
Global positioning system (GPS)

Related reference
Keys

This organization of links within a topic gives users a mental picture of what's important without overwhelming them with information or leaving them stranded without immediate possibilities for more information.

You can also organize a navigation pane (or a table of contents for a user's guide) by assembling these topics. The tasks are in a sequence that reflects the likely order in which users would do these tasks. You could keep these types of topics separate even in the navigation pane, labeling them as "Tasks," "Concepts," and "Reference." However, users might not think in these terms and so might not know where to look for the information that they want.

A more useful order might be to present the tasks at the top level and insert the concept and reference topics where they first apply, as follows:

Original

Installing InfoGPS

Turning on InfoGPS
  Global positioning system (GPS)
  Keys

Setting options

Setting the date and time

Orienting your InfoGPS
  Restrictions

Creating a route
  Components of a route

Following a route

Changing a route

Deleting a route

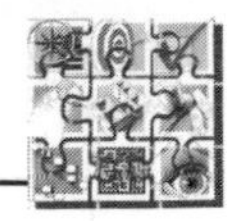

This list of contents uses all of the topics, but it has single entries under some tasks, such as "Components of a route" under "Creating a route." These single entries indicate a problem in the structure. Also, the starting tasks would be easier to find if they were grouped under a heading such as "Getting started." Users do these tasks when they start using the product and thereafter do them infrequently, or users might not need help after the first time that they do a getting-started task.

Adding a getting-started task as a *container* for other tasks might prompt a need for more topics such as an overview of the product. Such an overview would be a concept, and it could serve as a container for the single topics (both concept and reference topics) that were further down in the structure. Another new higher-level task might become a similar container for the tasks that deal with routes:

Revision

Installing InfoGPS

Getting started

Overview of InfoGPS

- ❑ Global positioning system (GPS)
- ❑ Components of a route
- ❑ Restrictions
- ❑ Keys

Turning on InfoGPS
Setting options
Setting the date and time
Orienting your InfoGPS

Using InfoGPS to navigate

Creating a route
Following a route
Changing a route
Deleting a route

Although the concept and reference topics might come early in this table of contents, the tasks can link to any of these topics as needed.

The following guidelines for organization, retrievability, and visual effectiveness are the guidelines that are most relevant to writing for the Web:

Quality characteristic	Relevant guidelines for writing for the Web
Organization	Organize information into discrete topics by type. Organize information consistently. Provide an appropriate number of subentries for each branch. Emphasize main points; subordinate secondary points. Reveal how the pieces fit together.
Retrievability	Facilitate navigation and search. Provide helpful entry points. Design helpful links.
Visual effectiveness	Use visual elements logically and consistently.

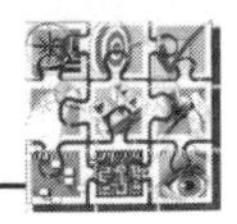

Revising technical information

Your first approach with information that you're revising might be to consider whether the information is needed at all. Think about the guidelines for task orientation and completeness in particular. Maybe some information doesn't belong, as in the task topic on page 332 that offered two sets of instructions for two different audiences. Maybe a different viewpoint on the information is needed. For example, as with the reference information on page 340, you might need to clarify the type of information and then present the information as appropriate for that type.

Maybe you need first to "translate" the technical jargon into something more comprehensible. This translation is like working from product specifications, as in the conceptual information on page 337. After you understand the information, you can decide how best to revise it.

In addition, you need to consider the needs of the particular audience, especially if it is an international audience, and the opportunities or constraints of the particular medium, such as the Web.

The process of revising requires you to:

1. Understand the information.
2. Improve the information.
3. Improve the information some more.

You can use the quality characteristics to approach technical information systematically so that you can more easily write it and revise it.

Chapter 12

Reviewing, testing, and evaluating technical information

Reviewing, testing, and evaluating technical information can take several forms, depending on who the reviewer is. This chapter deals with the role of the following kinds of reviewers:

- Technical reviewers
- Actual users or people similar to them
- Testers (who might also be writers, editors, visual designers, or human factors engineers)
- Editors (who might also be writers)
- Visual designers

Every writer needs reviewers—other people to see and use the information. Reviewers bring their own skills and perspectives to the information.

Just as reviewers have special skills, they also have different areas of focus. Not every reviewer needs to look for all of the same items. For example, technical reviewers need not comment on style, because that is a domain for editors. See "Who checks which quality characteristics?" on page 391 for a table that summarizes what the various kinds of reviewers should look for.

Not every project requires the same level of reviewing, testing, and evaluating. Some projects have few quality problems or low visibility, for example, and so do not warrant intensive efforts by many people. However, in general, the fewer resources spent on reviewing, testing, and evaluating the information for a project, the greater the risk of poor quality.

Inspecting technical information

A technical reviewer is an expert in the subject that the technical information covers. In general, technical reviewers are best qualified to evaluate the accuracy, completeness, concreteness, and visual effectiveness of technical information. However, technical reviewers might want more technical precision in the information than a particular audience needs.

The technical reviewer's role is to:

- Read the information critically, or use it, or both.
- Find problems with the information.
- Report the problems in a way that helps you understand the problems and fix them.

To help reviewers provide useful comments, you can talk with them about their role before a review and discuss the following advice about reading technical information, finding problems, and reporting them. You can also provide that information in your cover note to the technical reviewers when you announce a review. You can give them examples of the kinds of problems to look for (based on the pertinent guidelines in Table 3, "Quality characteristics for technical reviewers to focus on," on page 359) and of helpful kinds of review comments.

Reading and using the information

Reviewers need to put themselves in the place of the user as they read or use technical information. When you review information, you should actively question what you read and continually ask yourself: "If I were the user, could I find, understand, and use this information?"

Reviewers tend to assume that because they understand something, a user will also. However, this assumption is warranted only if the skills and situation of a reviewer are very like those of a typical user.

Reviewers who review online information need to look at and use the information online, not just in a printed form. Visual elements of the information, for example, might have more impact online.

Most important, reviewers should use the product while reviewing the information. For example, they can use a wizard to do a task when they consider how well the information supports the task. Ask reviewers to look at the information in the context of using the product.

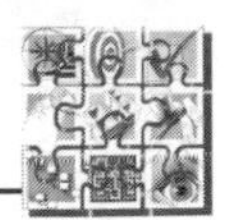

Finding problems

Table 3 shows the quality characteristics and guidelines that are most likely to benefit from scrutiny by technical reviewers.

Table 3. Quality characteristics for technical reviewers to focus on

Quality characteristic		Guideline
Easy to use	Task orientation	The information is appropriate for the intended audience.
		The focus is on real tasks.
		Step-by-step instructions are clear.
	Accuracy	The information has been verified.
		The information reflects the current subject or product.
		The information about a subject is consistent.
	Completeness	Only needed topics are covered.
		Each topic has just the detail that users need.
Easy to understand	Clarity	The focus is on the meaning.
	Concreteness	The examples are appropriate for the audience and subject.
		The examples are focused, realistic, accurate, and up to date.
		The code examples are easy to adapt.
		Unfamiliar information is related to familiar information.
Easy to find	Organization	Emphasis and subordination are appropriate.
	Retrievability	The index is complete.
		Links and cross-references are helpful and appropriate.
	Visual effectiveness	The graphics are meaningful and appropriate.
		The graphics complement the text.
		The visual cues help users find what they need.

Reporting problems

When technical reviewers report problems, comments like "I don't like it," "It's wrong," or "This needs fixing" are not much help. Ask them to be specific about what's wrong and to suggest a solution, as in:

- "This section has too much detail. Users need only the first two paragraphs."
- "All message identifiers have eight characters. Change to eight throughout."
- "Aligning the columns in this code as I've shown is what most users expect."

Even problems of accuracy and completeness can have more than one fix that will work. Ask reviewers to give enough information so that you can understand the reasoning behind a comment and discern the best fix. Applying the guidelines from the earlier chapters in this book can help you advise reviewers how to be specific about typical problems and how to suggest solutions.

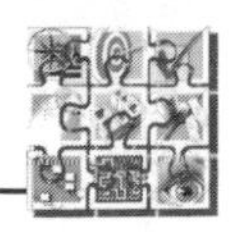

Testing information for usability

One way that you can improve and validate the quality of your information is to test it for usability. Through usability testing, you can find problems especially in task orientation and concreteness, as shown in Table 4. Users can provide direct and indirect feedback on the quality characteristics.

Table 4. Quality characteristics for users to focus on (Part 1 of 2)

Quality characteristic		Guideline
Easy to use	Task orientation	The information is appropriate for the intended audience.
		The information is presented from the user's point of view.
		A practical reason for information is evident.
		The focus is on real tasks.
		The headings reveal the tasks.
		The tasks are divided into discrete subtasks.
		Step-by-step instructions are clear.
	Accuracy	The information reflects the current subject or product.
		The information about a subject is consistent.
		The references to related information are correct.
	Completeness	Only needed topics are covered.
		Each topic has just the detail that users need.
		Information is repeated only when needed.
Easy to understand	Clarity	The focus is on the meaning.
		The elements flow from one to another.
		The technical terms are necessary and appropriate.
		The new terms are defined.
	Concreteness	The examples are appropriate for the audience and subject.
		The examples are easy to find.
		The code examples are easy to adapt.
		Unfamiliar information is related to familiar information.
		General language is used appropriately.

Table 4. Quality characteristics for users to focus on (Part 2 of 2)

Quality characteristic		Guideline
Easy to find	Organization	Emphasis and subordination are appropriate.
	Retrievability	The index has predictable entries.
		Links and cross-references are helpful and appropriate.
	Visual effectiveness	The graphics are meaningful and appropriate.
		The graphics complement the text.
		Visual cues help users find what they need.
		All users can access the information.

Early testing is most valuable for pointing out problems while you can still make major changes. Flaws in a design, such as lack of information in the user interface, can be hard to remedy later in a project.

Prototyping

A prototype is a mock-up of a design, perhaps on paper (sometimes called "low tech" or "low fidelity"). A prototype of a user interface, for example, can show how the supporting information in the interface works in various user situations. You can use such a prototype to get feedback from potential users or from people similar to potential users. Such a prototype avoids problems with installing actual software and can be carried out early in a project.

Testing outside a usability laboratory

Some testing outside the usability laboratory can be done internally, without real users. Tests that involve people such as technical reviewers, writers, and usability specialists can yield valuable information and yet not require a lot of time to prepare and run. In a walkthrough, for example, the people who are involved in developing the information try to use the information in a realistic way, by performing procedures and following instructions. This type of testing yields information that you can use to fix problems.

You have several possibilities for testing information outside a usability laboratory with actual users (perhaps during a beta program), as shown in Table 5.

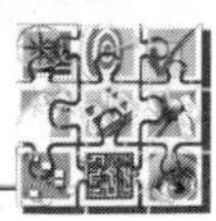

Table 5. Usability tests outside a usability laboratory

Type of test	What is tested	Procedure	Measurement
Field observation	Product and information as they are used in a real-world situation	Writers visit users and observe them using the product and information. The contextual inquiry form of observation also involves asking questions to make sure you are observing accurately and making correct inferences.	Qualitative data about what users do
Survey or questionnaire	Product and information	Users answer questions about the quality of the information and product.	Comparison of answers
Remote usability test	Navigation and content of online help and field-level help Product information with or without the product	Users participate in tests in one location, while test administrators in another location use special software to view the users' screens and listen through a conference call.	Successful completion of a task Time to do a task Number of errors
Instrumented beta test	Navigation and content of online help and field-level help	"Hooks" in the beta code enable recording which help topics or field-level help users access and in what circumstances. Programmers must provide the necessary hooks, and users must agree to this form of data collection.	Qualitative data about what users do
User edits or reviews	Information and the product in beta test	Users review the information and offer recommendations for improvement.	Periodic ratings of satisfaction on quality characteristics during the beta test

Using any of these forms of feedback, you can change the information to fix problems and improve the quality. When you are uncertain about a problem or how to fix it, you can return to the users to get more information or ask another set of users.

Testing in a usability laboratory

The usability laboratory offers the opportunity to:

- Observe a user and the user's work area through one or more cameras from another room
- Record the proceedings for later analysis
- Control the testing environment to limit interruptions

The usability laboratory makes it possible for more people to learn from the usability testing both while it is going on (through observation) and later (through the film or tape). A human factors engineer (someone who specializes in how people interact with things) can help you gather and analyze both qualitative and quantitative data. However, testing in a usability laboratory can also require a lot of time and resources.

Techniques for testing information in a usability laboratory are roughly grouped into evaluation tests and validation tests:

- *Evaluation tests* are done early in the development process to determine whether a design will meet the usability criteria if fully implemented.
- *Validation tests* are done late in the development process to determine whether the implementation meets the usability criteria.

The type of testing that you choose depends on your goals, time, and resources, and where you are in the development process.

Both evaluation tests and validation tests are done with actual users or with people whose experience and skills make them representative of users. These representative users can come from inside or outside the company. If they are from inside the company, they should not be familiar with the information.

Table 6 gives an overview of the procedures and measurements for both types of tests.

Table 6. Usability tests in a usability laboratory

Type of test	Procedure	Measurements
Evaluation tests	Users do scenario-based tests on the information, maybe without the product. Tests are generally iterative (test, improve, retest, improve, retest).	User comments Errors Ratings of satisfaction on quality characteristics such as task orientation, completeness, and clarity Comparisons of the preceding measures for competitive information
Validation tests	Users do scenario-based tests on the product and information together. The test team observes users; records actions, errors, and comments; later does statistical analysis on the data and summarizes the results.	Successful completion of a task Time to do a task Number of errors in retrieving information Ratings of satisfaction on quality characteristics such as task orientation, completeness, and clarity Comparisons of the preceding measures for competitive information

Evaluation tests

Compared with validation tests, evaluation tests tend to be less formal in these ways:

- The test participants and the test administrator interact more, in terms of asking questions and giving directions.
- Fewer scenarios are used, with fewer users for each.
- The measures are more subjective.

Testing with fewer users does not mean that your results are necessarily less valid. After even a few users, you will probably see the same errors in the same places. Showing that a problem exists can be enough to make a usability test worthwhile.

Use evaluation tests to achieve the following kinds of goals:

- To evaluate the overall organization of the information or components (such as topic titles, contents list, and examples)
- To determine which of two or more alternative designs is more usable
- To evaluate new information technologies (such as new search mechanisms)
- To compare the information for competitive products
- To compare a particular kind of information (such as tutorials) to that of the best-of-breed supplier of that type of information

Validation tests

Compared with evaluation tests, the test participants and the test administrator interact less, more scenarios are used, and the measures are more objective. Validation tests provide a good way to determine how your information compares with the information of a competitive product. You can, for example, compare satisfaction ratings, errors, time that was spent on a task, and successful completion of a task and thus gather benchmark measures.

You can use feedback from validation tests in these ways:

- To pinpoint and fix major usability problems before you make the information available
- To assess the interaction and flow of information with the product and whether users can get to the needed information
- To test whether the information helps users complete product tasks

- To validate quality characteristics of the information such as task orientation, completeness, and retrievability
- To get a final benchmark against the information of a competitive product

If you do not have enough time and resources to fix all of the problems, you can plan to resolve less serious usability problems through either of two means:

- A later release of the product
- Enhancements to the information that you make available on the Web

Validation tests are generally more difficult to design and take longer to conduct, although they yield more practical results. A validation test is probably appropriate in these cases:

- If the impact of your test will be wide ranging (for example, the information that you are testing will be used by several products or will be a template for other information)
- If you have a fairly stable product to test with
- If you have the time and resources

If you have done evaluation tests during design and development of the information and changed the information to deal with the problems that you found, the validation tests should not produce surprising results.

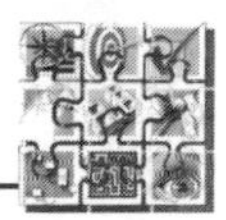

Testing technical information

Just as you cannot test quality into a product that has not been developed to meet quality standards, you cannot test quality into information that has been developed without regard for quality. However, appropriate testing can help you find and fix certain types of problems with information, as shown in Table 7.

Table 7. Quality characteristics for testers to focus on

Quality characteristic		Guideline
Easy to use	Task orientation	The focus is on real tasks.
		Step-by-step instructions are clear.
	Accuracy	The information has been verified.
		The information reflects the current subject or product.
		Corrections have been made based on tools.
		The references to related information are correct.
	Completeness	Each topic has just the detail that users need.
Easy to understand	Clarity	The language is unambiguous.
	Concreteness	The examples are focused, realistic, accurate, and up to date.
		The examples are easy to find.
		The context for examples and scenarios is described.
Easy to find	Organization	Organization of the tasks is by order of use.
		Organization of the topics enables quick retrieval.
	Retrievability	Navigation and search are easy.
		The index is complete.
		The table of contents has an appropriate level of detail.
		Linked-to information is easy to find in the target topic.
	Visual effectiveness	The textual information is legible.
		All users can access the information.

Using test tools

Writers should routinely test any kind of information by using a spelling checker. Be sure to look closely at the items that the checker finds. Do not assume, for example, that a strange word or acronym has been found because it is not in the checker's dictionary but should be; the item might be misspelled.

A grammar checker can be more time-consuming to use than a spelling checker, and so you will probably use one less often. However, such a checker can help you see the information in a new way. Getting a new perspective on the information is particularly important if you don't have an editor or a writing colleague to look at the information.

You can also use tools to test the following aspects of Web information:

- ❑ The links are correct.
- ❑ Appropriate tags are used to facilitate the use of screen readers.
- ❑ Tags are properly nested and will work in various browsers.
- ❑ A document that is formatted for online viewing opens correctly and has correct links.

You should also check that links are correct in any kind of online information, even if it doesn't appear on the Web.

Using test cases

To test a product, testers write and run test cases, which have steps for exercising parts of the product. These test cases include making sure that the product responds well when users make errors, such as when they enter the wrong format for a Web address.

Writers can create test cases to test parts of the online information such as information that should show up through using the search capability. For example, a test case might include several strings to search for, to ensure that the expected information is available and that the results are presented appropriately.

Writers can also run product test cases to make sure that the product information covers the items that are in the test cases. Given that many software development teams write inadequate design specifications, test cases can be a good source of information about how the product works. For example,

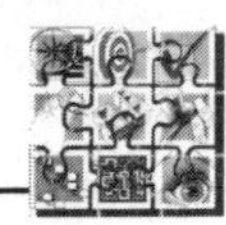

some test cases test that the software responds appropriately when users enter inappropriate data. The information should help users enter appropriate data and help them recover if they do not.

Testing the user interface

The writing team might have responsibility for at least some of the user interface, especially the interface for the product information. Testers (whether writers, editors, visual designers, or even human factors engineers) can check the interface in the following ways:

- Check the appearance of interface controls and icons, particularly for consistency of color and size and for legibility.
- Check the text on the screen and on interface controls, such as buttons.
- Test input methods, including the keyboard, mouse, and touch screen.
- Test navigation within and between windows, menus, or components.
- Test the expected behavior of interface controls and icons.
- Test whether appropriate messages inform users about problems or potential problems.

Editing and evaluating technical information

Conventional editing produces comments about the weaknesses in information. Using the quality characteristics, however, you can categorize both strengths and weaknesses. You can also see which quality characteristics of the information most need work and prioritize revisions. When you incorporate the quality characteristics into your editing, you probably need to modify your approach to editing.

Editors look at all aspects of the quality characteristics as they edit technical information. However, to verify the accuracy of the information, editors rely heavily on technical reviewers who are more familiar with the technical details.

If you are editing single source information that might be assembled in various ways, you have several choices for organizing your editing, particularly if more than one writer is working on the project:

- Edit a main task and all of its related task, concept, and reference information. With this approach you can determine whether the information is complete and whether links are appropriate.
- Edit all topics of a particular type, such as all concept topics. You can easily check for compliance with the guidelines for that type of topic.
- Edit the information after it has been assembled into a final format, such as a Portable Document Format (PDF) file.
- Edit the information that one writer has written, before you edit the information that another writer has written. This approach is probably reasonable only if the amount of information from each writer is small.

You might combine some of these approaches, depending on the schedule and the needs of the material.

When you edit technical information, you can apply the quality characteristics throughout the following steps:

- Prepare to edit.
- Get an overview.
- Read and edit the information.
- Look for specific information.
- Summarize your findings.
- Confer with the writer.

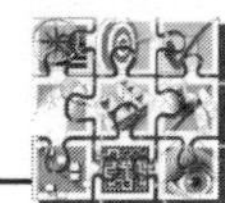

Preparing to edit

Familiarize yourself with the general plan for the information before you begin editing it. The writer should provide this information through appropriate parts of the documentation plan and other materials. You should find out the following information before you begin:

- ❑ Audience (including job tasks and experience level)
- ❑ Graphic design (including visual elements such as tabs and color)
- ❑ Delivery medium or media (such as Web, help, wizard, multimedia, and print)
- ❑ Expected number of topics and a hierarchical list of topics
- ❑ Planned changes in the content
- ❑ What style guidelines apply
- ❑ Plans for translation into other languages
- ❑ Codes for color and highlighting if you're editing only a printed form of online information

For printed information that isn't new, browse the latest edition if you have access to the book. You will probably see more clearly how the new edition is intended to differ from its predecessor.

The purpose of all this preparation is to develop an idea of what the finished information is meant to contain and look like. This preparation gives you a helpful set of expectations, and enables you to see what has and hasn't been realized in the information.

Before delivering a draft to you, the writer should use automated tools (such as a spelling checker and a grammar checker) and fix the problems that the tools found.

Getting an overview

When you begin editing information, survey it as a whole, as a user might.

Pay attention to the visual elements:

- ❑ Is the information presented in an attractive way?
- ❑ Are the text and graphical elements well balanced?
- ❑ Are tables and lists used effectively?

- ❑ Are the typefaces that are used for headings, text, and highlighting clear and helpful?
- ❑ Are graphical elements and typefaces used consistently?

For task information

Focus on task orientation by asking the following questions about the information:

- ❑ Is the intended audience clearly stated?
- ❑ Does the navigation pane or table of contents clearly show the real tasks for the product?
- ❑ Are the tasks presented in the order of use?
- ❑ Do all necessary tasks appear to be included?
- ❑ Are optional and conditional steps handled appropriately?
- ❑ Is task help accessible from the point where a user might need help?
- ❑ Can a user read the task help while doing the steps?
- ❑ What is the pattern of information for presenting tasks?
- ❑ Do task topics provide links or cross-references to supporting conceptual and reference information?

Checking the navigation pane or the table of contents (and the preface or the introduction for printed task information) can help you determine the task orientation, completeness, and organization of the information as a whole. For printed task information, you can also check whether all required parts of the front and back matter are in place.

For conceptual information

Focus on organization by asking the following questions about the information:

- ❑ Are major concepts presented as topics?
- ❑ Are concept topics related to appropriate task and reference topics?
- ❑ Are concept topics related to each other?
- ❑ Are graphics, examples, similes, and analogies used effectively in concept topics?
- ❑ Are minor concepts treated appropriately within other topics, such as with brief explanations and glossary definitions?

 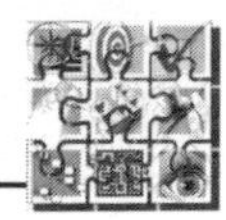

For reference information

Focus on retrievability and visual effectiveness by asking the following questions about the information:

- ❑ What retrievability aids does the information use? If the information is online, does it use color or graphics to draw the eye to topics of importance that recur (such as restrictions or operating system-dependent information)? If the information is printable or printed, does it use aids such as bleed tabs, in addition to the usual table of contents, index, and running feet or running heads?
- ❑ Does the organization enhance retrievability? Will the subdivisions of information in the table of contents seem logical to someone who is looking up information?
- ❑ Do the type style and format enhance retrievability? For example, do code examples stand out as examples? Is the explanation of a syntax element easy to pick out?
- ❑ Is the information retrievable by users who have impairments (such as blindness, color blindness, deafness, or mobility restrictions)?

You can also consider the organization and completeness of reference information by asking:

- ❑ If the information explains reference items such as classes, methods, commands, or programming statements, are the explanations in alphabetic order by name?
- ❑ If the information explains messages, are the explanations ordered by message identifier?
- ❑ If the information contains syntax diagrams, does it also have an explanation of how to read the diagrams?
- ❑ Are the headings appropriate for reference information? Gerund phrases indicate that task information has been mistakenly included or that reference topics are mistitled.
- ❑ Is the information presented in a pattern (such as purpose, syntax, restrictions, usage notes)?

Reading and editing the information

Read and edit the information in detail. For online information, establish an order for selecting and reading the information. For print information, this reading should be linear, from front to back. Defer more active testing, searching, and using the information until the next step.

After you get into the material, you will probably want to make notes of items to add to the style sheet and items to watch out for that you suspect might be a problem. At some point you can begin to list problems, perhaps under the quality characteristic where you think that they apply.

For task and concept information, keep these questions in mind:

- ❑ Can I understand the steps in a procedure?
- ❑ Do the procedures tell me what I need to know to do the task?
- ❑ Is the information easy to navigate online? Is the sequence clear?
- ❑ Can I understand topics the first time that I read them?
- ❑ Are new terms defined?
- ❑ Do I see anything that is obviously inconsistent or incorrect?
- ❑ Are the transitions from subject to subject logical?

For reference information, keep in mind that users will not read it from cover to cover, or even section by section. They will enter it almost anywhere, in search of small bits of information. So as you read and edit, make a list of the kinds of information that you come across—syntax diagrams, keyword explanations, restrictions, operating-system considerations, examples, and so on. You will use this list in the next step when you check retrievability.

Use the checklists at the end of each chapter in this book to help ensure that you look for the problems related to each quality characteristic. If you find more problems, add them to the lists where appropriate. Here's a brief list of major areas to check:

- ❑ Factual consistency
- ❑ Organizational consistency
- ❑ Appropriate level of detail
- ❑ Appropriate visual elements
- ❑ Consistent style

Check for factual consistency

Make sure that the information in examples, tables, and figures agrees with the surrounding text. Information about the same subject should be consistent.

Make sure that the keyword relationships shown in syntax diagrams agree with the explanations of the keywords. For example, a diagram might show that keyword X and keyword Y can coexist in the same command, but the explanations of X and Y might suggest that they are incompatible. Either the diagram or the explanations need to be fixed.

 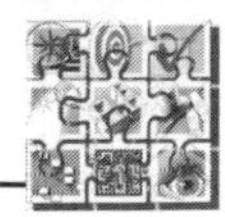

Also look at the examples and ensure that they can actually be derived from the syntax. Make sure that the examples and supporting text agree and that scenarios are consistent.

Check for organizational consistency

Expect to find repeated headings in help topics, especially those that describe commands or programming statements. This repetition is common for headings at lower levels. For example, an operator's reference might have these headings for every command: "Format," "Environments," "Keywords," "Usage," "Restrictions," and "Examples." Users would be confused if they found different headings under different commands, if some commands had all these headings and some had fewer, or if the sequence of headings varied from one command to another.

Even when a set of headings is used consistently, you should ensure that the same kind of information falls under a given heading, from one context to another. For example, the kind of information that falls under "Usage" in one context shouldn't appear under "Restrictions" in another.

Check for appropriate level of detail

The need for some details depends on the experience level of the audience. Other details are needed by any users and must be consistently provided. As you read the explanations of interface elements (fields, buttons, commands, statements, keywords), for example, make sure that their purpose is clearly stated. A similar requirement holds for examples: the purpose or effect of examples must be clear.

Check for appropriate visual elements

Looking at one type of visual element such as tables can reveal problems that you wouldn't otherwise notice. Consider, for example, whether the content of tables effectively supports or replaces text. If your audience might include vision-impaired users, check for one of the following methods as appropriate so that these users can still benefit from the information in graphics:

- ❑ The key content of graphics is summarized in surrounding text.
- ❑ For Web information, the graphic is briefly described in alternative text, or a longer description is provided as text that most users would not see.

Check for consistent style

Notice choices of style, and whether they are consistent throughout the information. Examples of style choices to check are terminology; highlighting; presentation of syntax diagrams; and parallelism of headings, list items, and index entries.

Looking for specific information

Check links or cross-references between topics to see whether information is linked appropriately. Look in the table of contents or index for topics that you have read; you should be able to find them again.

Check whether reference information has effective entry points to each of the kinds of information that you identified while reading the information. Give special attention to the index: every item on your list should be represented by many index entries, including synonyms.

This search gives you an idea of how useful, accurate, and complete the retrievability devices are.

Summarizing your findings

Use the checklist at the end of each chapter in this book to summarize the kinds of problems that you found.

If a visual designer reviewed the information, combine the comments from that review with your own before you make your final assessment. See "Reviewing the visual elements" on page 379 for relevant guidelines.

Your comments in the draft and the summary tell writers and planners about strengths that you find as well as changes that can improve the quality of the information. Your summary should identify the strengths and the weaknesses that, in your opinion, would have the most impact on how satisfied users are with the information.

Assigning problems to the quality characteristics

Some problems affect more than one quality characteristic. In that case, consider the importance of the problem to the user. When a problem might affect any of two or more quality characteristics, give preference to the quality characteristic that is more important to the user. For example, a typo like

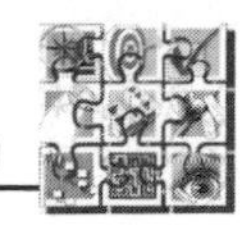

its instead of *it's* might be a style problem in one situation but an accuracy problem in another situation. Transposing letters in a file name has a definite impact on accuracy, but *form* instead of *from* might hardly be noticed.

However, also note when a problem affects another quality characteristic. For example, note when a problem that you report in concreteness also has an effect on retrievability. This information helps the writer understand the impact of a problem.

Style has a dependence on other quality characteristics, because some characteristics provide a basis for style decisions. For example, a style guideline might be made because following it promotes clarity, visual effectiveness, or retrievability. Style in itself does not have a rationale other than the force of convention.

You might choose to put a deviation from the chosen style guidelines in another quality characteristic when *both* of the following statements are true:

- Failure to follow the style has a major impact on the other quality characteristic.
- The error occurs frequently. The writer seems to have forgotten the rationale or seems not to understand it.

You might find other "rules" for assigning problems to quality characteristics as you become more familiar with thinking about problems in terms of the quality characteristics and their effect on users.

Assigning quality ratings

After you categorize and summarize your comments, you can assign ratings to the quality characteristics. A quality rating is a numeric representation of the quality of information on a particular characteristic or on the whole set of quality characteristics. You can use this quality rating as an indicator of the progress of information either during its development cycle or from release to release.

The quality rating is the result of:

- Thoroughly editing the information
- Summarizing your editorial comments
- Rating each of the nine quality characteristics

You can use the "Quality checklist" on page 387 to summarize your ratings. To say how satisfied you are with each quality characteristic, review the strengths and weaknesses that you classified in a given quality characteristic and then consider the meanings of the scale points.

As you work with quality ratings, you will probably find situations where one problem could be enough to bring a quality characteristic down a point and where many small problems would not change a score. Again, you will probably derive some rules to help you decide what ratings to give. You might decide, for example, that when in doubt about a rating for a quality characteristic, pick the better rating.

However, assigning a quality rating is not like grading a math test, where you subtract points for errors. Assigning a quality rating is more like comparing the information to what you consider average and then figuring out whether the strengths and problems that you found make the information better or worse than average.

You need to take into account where the information is in the development cycle. For example, an early draft might not have all of the information, but the writer should indicate what is missing and where it will go; the effect of the missing information on the completeness rating for the information would be much smaller than in a final draft.

Finally, when you determine the accuracy quality rating for information, take into account your level of knowledge about the subject. You probably need to consider whether the technical reviewers reported errors in the information and, if so, the number of errors and their impact.

Conferring with the writer

Even though writers need reviewers (especially editors), few people have thick skins when their writing draws critical comments. If you meet with the writer to go over your comments, you can make sure that the writer understands your suggested changes and your reasons for them. You can also find out areas where your suggested changes might not be appropriate. After all, you and the writer are working together to create the best possible information for the users.

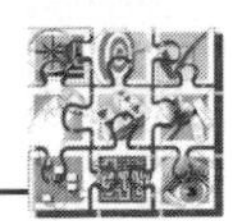

Reviewing the visual elements

Visual designers review information and software interfaces for how well their visual elements attract and motivate users and convey the intended meaning. Table 8 shows the relevant quality characteristics.

Table 8. Quality characteristics for visual designers to focus on

Quality characteristic		Guideline
Easy to use	Task orientation	The information is appropriate for the intended audience.
		The information is presented from the user's point of view.
		The focus is on real tasks.
	Accuracy	The information is accurate and technically current.
		The information about a subject is consistent.
	Completeness	Topics support users' tasks.
		Topics contain as much detail as users need.
		Patterns of information ensure proper coverage.
		The information is repeated only when necessary.
Easy to under-stand	Clarity	The focus is on the meaning.
		Similar topics are presented in a similar way.
	Concreteness	The examples are appropriate for the audience and subject.
		The examples are easy to find.
	Style	Style guidelines are followed.
Easy to find	Organization	Emphasis and subordination are appropriate.
		Users can see how the pieces fit together.
	Retrievability	Navigation and search are easy.
		The table of contents has an appropriate level of detail.
		Entry points are helpful.
		Linked-to information is easy to find in the target topic.
	Visual effectiveness	The graphics are meaningful and appropriate.
		The graphics complement the text.
		The visual elements are used for emphasis.
		The visual elements are logical and consistent.
		The visual elements are balanced.
		Visual cues help users find what they need.
		The textual elements are legible.
		All users can access the information.

When you review the visual elements, you can apply the quality characteristics throughout the following steps:

1. Prepare to review.
2. Get an overview.
3. Review individual visual elements.
4. Summarize your findings.
5. Confer with the editor or writer or both.

Preparing to review

Familiarize yourself with the plan for the information, and with the user interface if the information accompanies a software product. Get copies of the information design specifications and style guidelines that the writer used.

Before you begin your review, you should know the following information:

- ❑ Type of information (such as concept, task, or reference)
- ❑ Delivery medium or media (such as Web, help, wizard, or multimedia)
- ❑ Target audience, their tasks and experience level
- ❑ Visual design specifications and guidelines
- ❑ What style guidelines apply
- ❑ What changes are to be made
- ❑ Plans for translation into other languages
- ❑ Ways for the user to interact with the product

If your review is to happen along with an edit, schedule your review to follow the edit. If the review is online, determine whether you will enter your comments as part of the online review. If the editor used an online editing tool, ensure that your comments are differentiated, such as by using a different color for your comments. If the review is done on a printed copy, get the copy with the editor's comments, though you will not need to review these comments until later in the process.

Getting an overview

For your first pass through the information, look at the online or printed pages to get a first impression, without actually reading the text. Disregard the editor's comments. Step through the information, noting both positive and negative impacts of its presentation. You can supplement the visual effectiveness checklist on page 326 with the following list:

 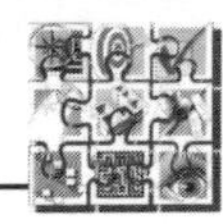

- ❑ Are pages visually balanced?
- ❑ Is white space used effectively? If the information is to be translated, can the white space accommodate the longer text that some languages require?
- ❑ Are task lists obvious? Is their sequence clear?
- ❑ Is the information free of visual clutter? Note, for example:
  - — Irrelevant graphics such as excessive screen captures
  - — Unnecessary lines and symbols
  - — Overabundant highlighting
  - — Distracting background patterns or animations
- ❑ Are the typeface and size that are used for text in each element legible and consistent?
- ❑ Are elements presented consistently and according to style guidelines? Check, for example, the following visual elements:
  - — Lists
  - — Tables
  - — Programming syntax
  - — Headings
  - — Highlighting
  - — Rules
  - — Spacing
- ❑ Are all illustrations rendered in a consistent style? Do the illustrations look as if the same illustrator did them? Check, for example, the following visual elements:
  - — Line characteristics
  - — Fonts and font sizes
  - — Color
  - — Shading and contrast
  - — Fill patterns
- ❑ Are the visual metaphors in the information the same as those in the product's user interface?
- ❑ If color, shading, or a patterned background is used, does it enliven and enhance the information without being overpowering or distracting? Are color choices logical and consistent? If text is placed on top of these graphic elements, is the text legible?
- ❑ Is the information broken into manageable chunks?
- ❑ Are visual retrievability aids used effectively? Check, for example, the following visual elements:

- — Tabs and labels
- — Rules and other separators
- — Captions
- — Icons
- — Color or shading
- — Running heads and running feet

- ❑ Is the organization and flow of the information clear at a glance?

When you review online information, consider also the following questions:

- ❑ Can visual elements be viewed in all supported resolutions?
- ❑ Are the visual elements usable in the supported resolutions?
- ❑ Are backgrounds simple and unobtrusive?
- ❑ Are links clearly identified and active?
- ❑ Are navigational aids consistent in design and placement?
- ❑ Are navigational aids obvious?
- ❑ Are graphics accompanied by alternate text that is meaningful for visually impaired users?

When you review multimedia presentations, consider also the following questions:

- ❑ Are all windows laid out consistently?
- ❑ Is a consistent, clear color scheme used throughout?
- ❑ Do visual transitions all use the same one or two methods?
- ❑ Is text sufficient without graphics or audio?

Reviewing individual visual elements

Go through the information again from the beginning, focusing on individual visual elements that you feel need attention. This time you might read the text that surrounds graphics and illustrations, especially if the text has been revised.

If an edit was done, review the editor's comments on visual elements and design. Note whether you agree and, if you disagree, why. At places where the editor has suggested adding or changing a visual element, read the surrounding text and make recommendations, if possible, on the design of the visual element.

Consider these issues as you focus on individual elements:

- ❑ Are illustrations meaningful and appropriate? Have the graphics been updated to match the revised text?
- ❑ Are the graphics legible and clear in all presentation media and formats?

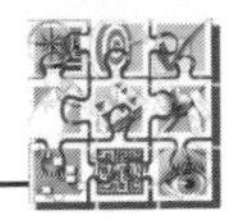

- In icons:
  - Are the visual metaphors that are used in the icons unambiguous and meaningful? If the information will be translated, are the icons universal, so that people of other cultures will understand them?
  - If you have seen these icons before, is their meaning here consistent with their earlier use?
  - If the icons show different states (such as running and stalled), is the meaning of each state clear?
- In images and toolbars:
  - Are the images that are used for toolbar actions clear?
  - Are common images used for common actions (for example, scissors for "cut")?
  - Are the images universal and not offensive to other cultures?
  - Is the meaning of toolbar buttons intuitive? Are they supplemented with tooltips?
  - Are images accompanied by descriptive text so that visually impaired users can understand them?
- In animations or video clips:
  - Are the animations smooth?
  - If the animations continuously loop, can users stop them? Can users easily figure out how to stop them?
  - Can users rerun animations that run once and stop? Can users easily figure out how to rerun them?
  - If the animations include text, can the text be read easily? Is the text presented long enough for users to read it? Is any text displayed intermittently, so that it blinks or flashes? Such text can be both annoying and unreadable.

You might find other visual elements that need a focused review, such as the user interface in general.

Summarizing your findings

Gather your significant concerns and comments into a report. Note the quality characteristic that you think is most seriously affected by each problem. The ultimate impact of a visual effectiveness problem is often to one of the other quality characteristics described in this book, such as clarity, accuracy, or organization.

If you identified areas where you disagree with the editor, mention these in your report. If you prefer, you can talk with the editor to resolve the differences in advance.

As much as possible, recommend solutions to problems that you identified. It helps to be familiar with the tools that were used to develop the information and its visual components. You can then recommend solutions that are within the capabilities of the tools.

Conferring with the editor or writer or both

If you did your review as part of an edit, give your comments and summary to the editor to compile into a summary report.

Schedule a meeting with the writer or both the writer and the editor to discuss the results of your visual review. If an illustrator or another designer was involved in developing the visual elements, include that person in the meeting, too. At the meeting, reach agreement among all participants on the changes that will be made in the visual elements.

Part 5

Appendixes

These appendixes bring together all of the quality characteristics for you to use in several ways. The checklist in Appendix A incorporates all of the guidelines from the chapters on the quality characteristics so that you can give a quality rating to technical information. The table in Appendix B shows which characteristics are appropriate for technical reviewers, users, testers, technical editors, visual designers, and peer editors to check. The material in Appendix C can help you understand how the quality characteristics apply to the various elements of information, from small to large and from syntactic to semantic.

Appendix A

Quality checklist

After you edit or review technical information, you can use this quality checklist to summarize your ratings for each quality characteristic. If you used the checklist at the end of each chapter to keep track of your findings, you can pull together your ratings here. You can then get an overall picture of the strengths and weaknesses of the information and make a plan for working on the weaknesses.

Quality checklist (Part 1 of 3)

Quality characteristic and guidelines	Quality rating
Easy to use	
Task orientation (page 17)	1 2 3 4 5
The information is appropriate for the intended audience.	1 2 3 4 5
The information is presented from the user's point of view.	1 2 3 4 5
A practical reason for the information is evident.	1 2 3 4 5
The focus is on real tasks.	1 2 3 4 5
The headings reveal the tasks.	1 2 3 4 5
The tasks are divided into discrete subtasks.	1 2 3 4 5
The step-by-step instructions are clear.	1 2 3 4 5
Accuracy (page 47)	**1 2 3 4 5**
The information has been verified.	1 2 3 4 5
The information reflects the current subject or product.	1 2 3 4 5
The information about a subject is consistent.	1 2 3 4 5
Corrections have been made based on findings from accuracy-checking tools.	1 2 3 4 5
The references to related information are correct.	1 2 3 4 5
Completeness (page 75)	**1 2 3 4 5**
All topics that support users' tasks are covered, and only those topics.	1 2 3 4 5
Each topic has just the detail that users need.	1 2 3 4 5
Patterns of information ensure proper coverage.	1 2 3 4 5
Information is repeated only when needed.	1 2 3 4 5
Easy to understand	
Clarity (page 103)	**1 2 3 4 5**
The focus is on the meaning.	1 2 3 4 5
The language is unambiguous.	1 2 3 4 5
The elements are short.	1 2 3 4 5
The elements flow from one to another.	1 2 3 4 5

Legend:

1—Very satisfied: Among the best; could be used as a model.
2—Satisfied: Solid, professional work.
3—Neither satisfied nor dissatisfied: OK—par for the course; overall, neither praiseworthy nor blameworthy.
4—Dissatisfied: Clearly subpar.
5—Very dissatisfied: Among the worst.

Quality checklist (Part 2 of 3)

Quality characteristic and guidelines	Quality rating
Similar information is presented in a similar way.	1 2 3 4 5
The technical terms are necessary and appropriate.	1 2 3 4 5
Each term that is new to the intended users is defined.	1 2 3 4 5
Concreteness (page 149)	**1 2 3 4 5**
The examples are appropriate for the audience and subject.	1 2 3 4 5
The examples are focused, realistic, accurate, and up to date.	1 2 3 4 5
The examples are easy to find.	1 2 3 4 5
The code examples are easy to adapt.	1 2 3 4 5
Scenarios illustrate tasks and provide product overviews.	1 2 3 4 5
The context for examples and scenarios is described.	1 2 3 4 5
Unfamiliar information is related to familiar information.	1 2 3 4 5
General language is used appropriately.	1 2 3 4 5
Style (page 181)	**1 2 3 4 5**
Grammar is correct.	1 2 3 4 5
Spelling is correct and consistent.	1 2 3 4 5
Punctuation is consistent and appropriate.	1 2 3 4 5
The tone is appropriate.	1 2 3 4 5
The style is active.	1 2 3 4 5
The mood is appropriate.	1 2 3 4 5
Templates and boilerplate text are implemented appropriately.	1 2 3 4 5
Style guidelines are followed.	1 2 3 4 5
Easy to find	
Organization (page 215)	**1 2 3 4 5**
The information is organized into discrete topics by type.	1 2 3 4 5
Organization of the tasks is by order of use.	1 2 3 4 5
Organization of the topics enables quick retrieval.	1 2 3 4 5
Contextual information is separated from other types.	1 2 3 4 5

Legend:

1—Very satisfied: Among the best; could be used as a model.
2—Satisfied: Solid, professional work.
3—Neither satisfied nor dissatisfied: OK—par for the course; overall, neither praiseworthy nor blameworthy.
4—Dissatisfied: Clearly subpar.
5—Very dissatisfied: Among the worst.

Quality checklist (Part 3 of 3)

Quality characteristic and guidelines	Quality rating
Organization of the information is consistent.	1 2 3 4 5
Each branch has an appropriate number of subentries.	1 2 3 4 5
Main points are emphasized; secondary points are subordinated.	1 2 3 4 5
Users can see how the pieces fit together.	1 2 3 4 5
Retrievability (page 245)	**1 2 3 4 5**
Navigation and search are easy.	1 2 3 4 5
The index is complete and consistent.	1 2 3 4 5
The table of contents has an appropriate level of detail.	1 2 3 4 5
Entry points are helpful.	1 2 3 4 5
Links are appropriate.	1 2 3 4 5
Links are helpful.	1 2 3 4 5
Linked-to information is easy to find in the target topic.	1 2 3 4 5
Visual effectiveness (page 277)	**1 2 3 4 5**
The graphics are meaningful and appropriate.	1 2 3 4 5
The graphics complement the text.	1 2 3 4 5
The visual elements are used for emphasis.	1 2 3 4 5
The visual elements are logical and consistent.	1 2 3 4 5
The visual elements are balanced in number and placement.	1 2 3 4 5
Visual cues help users find what they need.	1 2 3 4 5
The textual elements are legible.	1 2 3 4 5
The color and shading are discreet and significant.	1 2 3 4 5
All users can access the information.	1 2 3 4 5

Legend:

1—Very satisfied: Among the best; could be used as a model.

2—Satisfied: Solid, professional work.

3—Neither satisfied nor dissatisfied: OK—par for the course; overall, neither praiseworthy nor blameworthy.

4—Dissatisfied: Clearly subpar.

5—Very dissatisfied: Among the worst.

Appendix B

Who checks which quality characteristics?

Writers have responsibility for checking all of the items that affect the quality of technical information. Technical editors have a similar responsibility except that they usually are not responsible for verifying the accuracy of the information. Peer editors (or copy editors in some situations) typically edit near the end of the development cycle; they need to check certain aspects of all of the quality characteristics.

The more limited responsibility of technical reviewers (people skilled in the content of the information) spans several quality characteristics—mainly accuracy, completeness, concreteness, and visual effectiveness. Sometimes technical reviewers are asked to check certain aspects of clarity, organization, and retrievability.

During usability testing, users (or people similar to users) provide information that mainly affects task orientation but also parts of all the other quality characteristics.

The work of testers mainly affects accuracy and concreteness.

In addition to checking all aspects of visual effectiveness, visual designers need to check certain aspects of clarity, concreteness, style, and retrievability.

Who checks which quality characteristics (Part 1 of 4)

Quality characteristic and guidelines	Technical reviewers	Users	Testers	Technical editors	Visual designers	Peer editors
Easy to use						
Task orientation						
The information is appropriate for the intended audience.	✓	✓		✓	✓	
The information is presented from the user's point of view.		✓		✓	✓	
A practical reason for the information is evident.		✓		✓		
The focus is on real tasks.	✓	✓	✓	✓	✓	
The headings reveal the tasks.		✓		✓		✓
The tasks are divided into discrete subtasks.		✓		✓		✓
The step-by-step instructions are clear.	✓	✓	✓	✓		✓
Accuracy						
The information has been verified.	✓		✓			
The information reflects the current subject or product.	✓	✓	✓	✓		✓
The information about a subject is consistent.	✓	✓		✓		✓
Corrections have been made based on findings from accuracy-checking tools.			✓	✓		✓
The references to related information are correct.		✓	✓	✓		✓
Completeness						
All topics that support users' tasks are covered, and only those topics.						
Each topic has just the detail that users need.	✓	✓	✓	✓		
Patterns of information ensure proper coverage.				✓		✓
Information is repeated only when needed.		✓		✓		

Who checks which quality characteristics (Part 2 of 4)

Quality characteristic and guidelines	Technical reviewers	Users	Testers	Technical editors	Visual designers	Peer editors
Easy to understand						
Clarity						
The focus is on the meaning.	✓			✓	✓	✓
The language is unambiguous.		✓	✓	✓		✓
The elements are short.				✓		✓
The elements flow from one to another.		✓		✓		
Similar information is presented in a similar way.				✓	✓	
The technical terms are necessary and appropriate.		✓		✓		
Each term that is new to the intended users is defined.		✓		✓		✓
Concreteness						
The examples are appropriate for the audience and subject.	✓	✓		✓	✓	
The examples are focused, realistic, accurate, and up to date.	✓		✓	✓		✓
The examples are easy to find.		✓	✓	✓	✓	✓
The code examples are easy to adapt.	✓	✓		✓		✓
Scenarios illustrate tasks and provide overviews.				✓		✓
The context for examples and scenarios is described.			✓	✓		✓
Unfamiliar information is related to familiar information.	✓	✓		✓		
General language is used appropriately.		✓		✓		

Who checks which quality characteristics (Part 3 of 4)

Quality characteristic and guidelines	Technical reviewers	Users	Testers	Technical editors	Visual designers	Peer editors
Style						
Grammar is correct.				✓		✓
Spelling is correct and consistent.				✓		✓
Punctuation is consistent and appropriate.				✓		✓
The tone is appropriate.		✓		✓		✓
The style is active.				✓		✓
The mood is appropriate.				✓		✓
Templates and boilerplate text are implemented appropriately.				✓		✓
Style guidelines are followed.				✓	✓	✓
Easy to find						
Organization						
The information is organized into discrete topics by type.		✓		✓		
Organization of the tasks is by order of use.		✓	✓	✓		
Organization of the topics enables quick retrieval.		✓	✓	✓		
Contextual information is separated from other types.				✓		
Organization of the information is consistent.		✓		✓		
Each branch has an appropriate number of subentries.				✓		✓
Main points are emphasized; secondary points are subordinated.	✓			✓	✓	
Users can see how the pieces fit together.		✓		✓	✓	

Who checks which quality characteristics (Part 4 of 4)

Quality characteristic and guidelines	Technical reviewers	Users	Testers	Technical editors	Visual designers	Peer editors
Retrievability						
Navigation and search are easy.		✓	✓	✓	✓	
The index is complete and consistent.		✓	✓	✓		✓
The table of contents has an appropriate level of detail.		✓	✓	✓	✓	✓
Entry points are helpful.		✓		✓	✓	
Links are appropriate.	✓	✓		✓		
Links are helpful.		✓		✓		
Linked-to information is easy to find in the target topic.		✓	✓	✓	✓	
Visual effectiveness						
The graphics are meaningful and appropriate.	✓	✓		✓	✓	
The graphics complement the text.	✓			✓	✓	✓
The visual elements are used for emphasis.				✓	✓	
The visual elements are logical and consistent.				✓	✓	✓
The visual elements are balanced in number and placement.				✓	✓	
Visual cues help users find what they need.	✓	✓		✓	✓	✓
The textual elements are legible.			✓	✓	✓	
The color and shading are discreet and significant.				✓	✓	
All users can access the information.		✓	✓	✓	✓	✓

Appendix C

Quality characteristics and elements

Many technical writing books are organized around information elements. They deal with what to do and not do at the level of words, sentences, tables, and so on.

Table 9 on page 399 shows how the quality characteristics apply to these elements.

Looking at the quality characteristics

Some quality characteristics operate mainly from the bottom up. Large semantic elements (such as task topics) have the quality characteristic because the small elements (such as headings and ordered lists) do. You need to focus on the small elements to ensure that information has these quality characteristics:

- ❑ Accuracy
- ❑ Clarity
- ❑ Concreteness

Some quality characteristics operate more from the top down. You need to consider the large semantic elements to ensure that information has these quality characteristics:

- ❑ Completeness
- ❑ Organization
- ❑ Task orientation

Some quality characteristics are pervasive. They apply to every element. You can use various techniques at all levels of elements to enhance these quality characteristics:

- ❑ Retrievability
- ❑ Style
- ❑ Visual effectiveness

Looking at the elements

Few elements participate in all the characteristics. Examples are a notable exception. Their widespread participation in the quality characteristics underlines their importance.

You can recognize semantic elements by their content, and syntactic elements by their form.

In Table 9, the following symbols are used:

✓= A characteristic that is especially important for the given element.
x = A characteristic that is important for the given element.

Table 9. Quality characteristics and elements (Part 1 of 2)

Structure (from syntactic elements to semantic elements)	Quality characteristics								
	Task orientation	Accuracy	Completeness	Clarity	Concreteness	Style	Organization	Retrievability	Visual effectiveness
Small elements—mainly syntactic									
Word, phrase		x		✓	x	x		x	
Clause, sentence		x		✓	x	x		x	
Paragraph	x	x	x	x		x	✓	x	x
Heading, label	✓	x		x	x	x	x	✓	x
List, ordered	✓	x	x	✓	x	x		x	x
List, unordered	x	x	x	✓	x	x		x	x
Table	x	x	x	x	x	x		✓	x
Figure, illustration, screen capture		x	x	x	✓	x		x	✓
Larger elements—mainly from book paradigm									
Title, edition notice, trademarks		✓	✓			x			
Table of contents	x	x	x		x	x	x	✓	x
Overview, summary		x	x	x	x	x	✓	✓	x
Cross-reference, link, or citation		✓	x	x	x	x	✓	✓	x
Glossary		x	✓	x	x	x		x	x
Resources		x	x			x		✓	x
Index	x	x	✓	x		x		✓	x

Table 9. Quality characteristics and elements (Part 2 of 2)

Structure (from syntactic elements to semantic elements)	**Quality characteristics**								
	Task orientation	Accuracy	Completeness	Clarity	Concreteness	Style	Organization	Retrievability	Visual effectiveness
Variable elements—mainly semantic									
Example	x	x	x	x	✓	x	x	x	x
Analogy, simile		x		x	✓	x			
Scenario	✓	x	x	x	✓	x			x
Term, definition, concept		x	x	✓	x	x		x	x
Instruction, guideline, hint, wizard, cue card	✓	x	x	x	x	x	x	x	x
Message	✓	x	x	x	x	x		x	
Large semantic elements									
Task topic	✓		x			x	x	✓	x
Concept topic	✓			x	✓	x		✓	x
Reference topic		✓	x		x	x	x	✓	x
Contextual topic	x	✓	x	x	x	x		✓	x
Task and concept composite (such as guide, tutorial)	✓		x			x	x	✓	x
Reference composite		✓	x		x	x	x	✓	x

Resources and references

Easy to use

Beyer, Hugh, and Karen Holtzblatt. *Contextual Design: Defining Customer-Centered Systems*. San Francisco: Morgan Kaufmann Publishers, 1998.

Contextual design relies on forming an accurate understanding of how people work, their environment, the things and the language that they use to do their work, and even the culture and policies that affect their work. This book provides techniques for interviewing and observing users, interpreting the data, and applying the results to designing a system that meets users' needs.

Carroll, John M. *Minimalism beyond the Nurnberg Funnel*. Cambridge, MA: The MIT Press, 1998.

Both practitioners and academicians wrote papers for this book to take stock of minimalism. Three papers are authored or coauthored by John Carroll. In one article, he presents four minimalist principles: choose an action-oriented approach; anchor the tool in the task domain; support error recognition and recovery; and support reading to do, study, and locate.

Carroll, John M. *The Nurnberg Funnel: Designing Minimalist Instruction for Practical Computer Skill*. Cambridge, MA: The MIT Press, 1990.

Carroll, whose usability research at IBM in the early 1980s first showed the efficacy of minimalism for training materials, presents an approach that supports the natural tendency of many new users to learn through trial and error.

Hackos, JoAnn T., and Janice C. Redish. *User and Task Analysis for Interface Design*. New York: John Wiley & Sons, 1998.

This in-depth treatment of user and task analysis includes how to apply the analysis results to documentation and training.

Johnson, Jeff. *GUI Bloopers: Don'ts and Do's for Software Developers and Web Designers*. San Francisco: Morgan Kaufmann Publishers, 2000.

This book presents eight principles such as "Focus on the users and their tasks, not the technology" and "Conform to the users' view of the task" and then presents many examples of problems with graphical user interfaces.

Mandel, Theo. "Quality Technical Information: Paving the Way for Usable Print and Web Interface Design." *Journal of Computer Documentation.* Volume 26, Issue 3, August 2002.

In this commentary, Mandel evaluates *Producing Quality Technical Information*, which is the basis for the follow-on book, *Developing Quality Technical Information*. He observes that the aspects of quality that the book described almost 20 years ago can easily and usefully address key issues of interface design and usability for today's professional designers and developers.

Mayhew, Deborah J. *Principles and Guidelines in Software User Interface Design*. Upper Saddle River, NJ: Prentice Hall PTR, 1992.

This book provides practical assistance to interface, visual, and information designers about how to build usable software interfaces. Every chapter has both theoretical foundations and practical guidelines.

Microsoft Corporation. *Microsoft Windows User Experience.* Redmond, WA: Microsoft Press, 1999.
http://msdn.microsoft.com/library/default.asp?url=/library/en-us/dnwue/html/welcome.asp.

Microsoft espouses the following principles for designing user interfaces: user in control, directness, consistency, forgiveness, feedback, aesthetics, and simplicity. The chapter about user assistance indicates the typical audience for forms of user assistance such as status bar messages, reference help, and wizards and briefly discusses each form.

Schriver, Karen A. *Dynamics in Document Design: Creating Text for Readers*. New York: John Wiley & Sons, 1996.

This book focuses on document design as bringing together words and pictures in a clear, organized way to benefit readers. Schriver discusses three methods of audience analysis and how they can work together: classification (audience demographics), intuition (imagined readers), and feedback (audience participation in design). Schriver suggests ways to go from an audience analysis to the information that the audience needs.

User Interface Engineering. "Docs in the Real World." www.uie.com/realdocs.htm.

This article describes the findings of usability consultants at User Interface Engineering when they observed real users as they used technical information in their jobs.

Spool, Jared M. Tara Scanlon, Will Schroeder, Carolyn Snyder, and Terry DeAngelo. *Web Site Usability: A Designer's Guide*. San Francisco: Morgan Kaufmann Publishers, 1998.

This book presents data and guidelines based on extensive usability testing of Web sites.

Easy to understand

Broadbent, D.E. "The Magic Number Seven after Fifteen Years." In A. Kennedy and A. Wilkes (eds.), *Studies in Long-Term Memory*. New York: John Wiley & Sons, 1975, 3-18.

This research questioned Miller's earlier finding about the human ability to remember seven items plus or minus two.

Brogan, John A. *Clear Technical Writing*. New York: McGraw-Hill, 1973.

This book is packed with examples and exercises for learning how to eliminate "gobbledygook" from your writing. It focuses on major sources of unclear writing, such as redundancies, weak verbs, abstract nouns, showy writing, and improper subordination.

The Chicago Manual of Style: The Essential Guide for Writers, Editors, and Publishers. 15th Edition. Chicago and London: University of Chicago Press, 2003.

This style guide is a widely used general reference on matters of academic and technical style. It has chapters devoted to topics such as punctuation, numbers, and indexes. The latest edition has added a chapter on grammar, which treats each part of speech, and on word usage, which includes an alphabetical list of troublesome words from a and an to your and you're. The book has also broadened its scope to nonbook and nonprint documents while still concentrating on scholarly books and journals. You can consult the Chicago Manual FAQ at www.press.uchicago.edu/Misc/Chicago/cmosfaq/cmosfaq.html.

Goldstein, Norm, editor. *The Associated Press Stylebook and Briefing on Media Law*. Cambridge, MA: Perseus Press, 2002.

The *AP Stylebook and Briefing on Media Law* is the primary style reference for many newspapers and magazines. In a single alphabetical list, it covers capitalization, punctuation, word usage, and other style topics. It gives good advice in many areas, but remember that the news media use slightly different conventions than technical publications do. This book has a section on Internet guidelines, which includes definitions of terms and advice for searching the Web.

Hacker, Diana. *A Writer's Reference*. 4th Edition. Boston, MA: Bedford/St. Martin's, 1999.

This handbook has chapters on composition and style, correctness, and format (including basic grammar).

Hale, Constance, and Jessie Scanlon. *Wired Style: Principles of English Usage in the Digital Age*. New York: Broadway Books, 1999.

Written by two editors at *Wired* magazine, most of this book is a glossary of terms, from *Abilene Project* to *zine*, associated with communication over computer networks. The ten principles, very briefly discussed, include "play with voice" (in the sense of tone), "capture the colloquial," and "go global."

LeCompte, D. "Seven, Plus or Minus Two, Is Too Much to Bear: Three (or Fewer) Is the Real Magic Number." *Proceedings of the Human Factors and Ergonomics Society*, 1999, 289-292.

This article reports the latest research on how much readers can remember.

Lynch, Patrick J., and Sarah Horton. *Web Style Guide: Basic Design Principles for Creating Web Sites*. Second Edition. New Haven: Yale University Press, 2002.

Based on the authors' experience with the Yale and Dartmouth Web sites, this style guide aims at improving the design of Web sites, including Web pages, graphics, and multimedia. This guide focuses on clarity and style, but also deals with organization, navigation, and visual effectiveness.

MacGregor, J.N. "Short-Term Memory Capacity: Limitation or Optimization?" *Psychological Review*, 1987, 94(1), 107-108.

This article presents another look at how much readers can remember.

Microsoft Corporation. *Microsoft Manual of Style for Technical Publications*. Redmond, WA: Microsoft Press, 1998.

This style guide for writers who document Microsoft products and technologies is arranged alphabetically. It has entries from "abbreviations and acronyms" to "zoom in, zoom out." This style guide often provides a rationale for its guidelines, emphasizing clarity and consistency, and can apply more broadly to the computer industry. This book is available for download at www.microsoft.com/downloads/details.aspx?displaylang=en&FamilyID=B494D46B-073F-46B0-B12F-39C8E870517A.

Miller, George A. "The Magical Number Seven, Plus or Minus Two: Some Limits on Our Capacity for Processing Information." *The Psychological Review*, 1956, 81-97.

This article reports the first research that led to guidelines about how much to expect users to remember. Later research indicates that these guidelines might be too high.

Sun Technical Publications. *Read Me First: A Style Guide for the Computer Industry*. Upper Saddle River, NJ: SunSoft Press/Prentice Hall, 1996.

This style guide is geared primarily to print documentation. However, its information about topics such as writing style, writing for an international audience, and using illustrations can help you develop a company style guide. The book also has a chapter about indexing a technical manual.

Tarutz, Judith A. *Technical Editing: The Practical Guide for Editors and Writers*. Cambridge, MA: Perseus Press, 1992.

This book includes guidelines for creating a style guide and for editing with various purposes: editing for accuracy, editing for style, editing for usability, and editing for an international audience. One chapter deals with the 100 most common errors, grouped in the following categories: fuzzy thinking; style and usage; grammar and syntax; technical accuracy; judgment, taste, and sensitivity; typography and graphics; legal and ethical concerns; and proofreading oversights.

Thrush, Emily. "Plain English: A Study of English Vocabulary and International Audiences." *Technical Communication*, August 2001, 48 (3), 289-96.

Williams, Joseph M. *Style: Ten Lessons in Clarity and Grace*. Sixth Edition. New York: Addison-Wesley Educational Publishers, Inc., 2000.

This book has a broader focus than just technical information, but its lessons about coherence, emphasis, and conciseness are valuable for technical writers. The book has many examples and an in-depth analysis of them.

World Wide Web Consortium. www.w3c.org.

This Web site includes many standards, among them XML, XSL, and accessibility.

Zinsser, William. *On Writing Well: The Classic Guide to Writing Nonfiction*. 25th Anniversary Edition. New York: Harper Resource, 2001.

While focusing on how to write good English, Zinsser offers principles (the transaction, simplicity, clutter, style, the audience, words, and usage) and methods (unity, the lead and the ending, bits and pieces). Primarily a journalist and a teacher of writing, Zinsser includes a chapter about writing for science and technology and several chapters about writing nonfiction.

Easy to find

Bonura, Larry S. *The Art of Indexing*. New York: John Wiley & Sons, 1994.

Bonura offers five criteria for a good index: accuracy, depth of indexing, conciseness, cross-references, and logical headings. Bonura also emphasizes knowing your audience. This book includes a sample style guide for indexing.

Horton, William. *The Icon Book: Visual Symbols for Computer Systems and Documentation*. New York: John Wiley & Sons, 1994.

This book is an excellent guide for designers of icons for graphical user interfaces. It includes extensive recommendations for things to consider when designing for specific audiences, including international and cultural considerations.

Lathrop, Lori. "Evaluating Index Usability." www.indexingskills.com/evalusab.pdf.

Lathrop, a professional indexer, provides 19 areas in which to evaluate an index, along with tips for improving an index in each area.

Lopuck, Lisa. *Designing Multimedia: A Visual Guide to Multimedia and Online Graphic Design*. Berkeley, CA: Peachpit Press, 1996.

Although the main focus of this book is multimedia, it offers a good set of guidelines for visual designers on the planning, designing, and development of online graphic media, including user interface design and design for the Web. It includes information about the technical hurdles of designing for multiple operating systems, hardware, software, and color palettes.

Lopuck, Lisa. *Web Design for Dummies*. New York: John Wiley & Sons, 2001.

This book includes practical advice for planning a Web site, designing the graphics, and testing the interface.

Nielsen, Jakob. Alertbox. www.useit.com/alertbox.

Since 1995, this site has offered monthly or bimonthly articles about Web topics, most of which are based on Nielsen's usability studies. His articles about fighting linkrot (14 June 1998), useful navigation (9 January 2000), usability metrics (21 January 2001), and reducing redundancy (9 June 2002) are especially useful for technical writers.

Pepper, Steve. "The TAO of Topic Maps: Finding the Way in the Age of Infoglut." *Proceedings of XML Europe 2000*. 12-16 June 2000. www.gca.org/papers/xmleurope2000/papers/s11-01.html.

This paper discusses the topic map standard, particularly topics, associations, and occurrences, along with identity, facets, and scope.

Pfeiffer, William S. *Technical Writing: A Practical Approach*. Fifth Edition. Prentice Hall, 2002.

This book has chapters about methods of organizing information, including induction and deduction.

National Cancer Institute. "Research-Based Web Design and Usability Guidelines." http://usability.gov/guidelines/.

This Web site offers guidelines in areas such as links, navigation, search, and accessibility and cites the research to support each guideline. The site also evaluates the strength of the evidence.

Rosenfeld, Louis, and Peter Morville. *Information Architecture for the World Wide Web*. Sebastopol, CA: O'Reilly & Associates, Inc., 1998.

This book emphasizes designing information from the user's perspective. It includes chapters about organizing complex content for delivery on the Web, and designing systems for navigation, labeling, and search.

Rowland, Marilyn, and Diane Brenner, editors. *Beyond Book Indexing*. Medford, NJ: Information Today, Inc., 2000.

This book features articles by indexing professionals primarily about indexing for the Web.

Toub, Steve. "How to Design a Table of Contents." *Web Techniques*. February 1999. www.newarchitectmag.com/archives/1999/02/desi/.

This article discusses guidelines for dealing with aspects of a table of contents for a Web site such as availability, balance, contrast, and organization.

User Interface Engineering. "As the Page Scrolls." www.uie.com/scrollin.htm.

This article describes the results of research into how users scroll on Web sites. It recommends providing longer pages, using content links rather than category links, and grouping links.

Weinman, Lynda. *Designing Web Graphics.3*. Third Edition. Indianapolis, IN: New Riders Publishing, 1999.

This book provides a wealth of information for creators of graphics for the Web. It provides technical guidance about how to work within the constraints of varying Web browsers, color palettes, and operating systems. Some guidance about good design of graphics, type, and page layout is included. Also included is a CD containing many sample files as well as several shareware programs to assist with Web graphic design.

Putting it all together

Dumas, J.S., and J.C. Redish. *A Practical Guide to Usability Testing*. Norwood, NJ: Ablex Publishing Corporation, 1994.

This book covers several methodologies for integrating usability testing into the development process. The focus is on overall product usability, including documentation.

Hackos, JoAnn T. *Content Management for Dynamic Web Delivery*. New York: John Wiley & Sons, 2002.

This book includes information about how to break linear content into more flexible modules, such as you would put into a content management system.

Hackos, JoAnn T. *Managing Your Documentation Projects*. New York: John Wiley & Sons, 1994.

This book explains how to build in quality while using a structured process of planning, designing, implementing, producing, and maintaining information. It discusses how to conduct reviews. The information is probably most useful for large projects.

Hackos, JoAnn T., and Dawn M. Stevens. *Standards for Online Communication: Publishing Information for the Internet/World Wide Web/Help Systems/Corporate Intranets*. New York: John Wiley & Sons, 1997.

Emphasizing information design, this book focuses on analyzing information needs (in terms of procedural, conceptual, reference, and instructional information), designing online systems, and implementing the design. It includes information about organizing topics; using retrievability techniques such as links, tables of contents, indexes, and search; writing for readability; and adding graphics.

Hammerich, Irene, and Claire Harrison. *Developing Online Content: The Principles of Writing and Editing for the Web*. New York: John Wiley & Sons, 2002.

This book includes a discussion of the four C's of quality Web content: credibility, clarity, conciseness, and coherence.

Hoft, Nancy L. *International Technical Communication: How to Export Information about High Technology*. New York: John Wiley & Sons, 1995.

This book includes useful information about detecting and fixing cultural bias, writing to optimize translatability, creating graphics and multimedia presentations that appeal to a global audience, and reviewing and testing source and target product information.

Horton, William. *Designing Web-Based Training: How to Teach Anyone Anything at Anytime*. New York: John Wiley & Sons, Inc., 2000.

This book presents guidelines for organizing learning sequences and motivating learners, some of which might be more broadly applicable to technical information outside Web-based training. The book also considers training across countries and cultures.

Krug, Steve. *Don't Make Me Think: A Common Sense Approach to Web Usability*. Indianapolis, IN: New Riders Publishing, 2000.

"Test early and often" is the mantra of this book, which includes techniques to avoid spending lots of money and time on usability testing and yet get lots of value.

Mayhew, Deborah K. *The Usability Engineering Lifecycle: A Practitioner's Handbook for User Interface Design*. San Francisco: Morgan Kaufmann Publishers, 1999.

This book shows how to integrate usability techniques throughout the software development process, including development for the Web.

McAlpine, Rachel. *Web Word Wizardry: A Guide to Writing for the Web and Intranet*. Berkeley, CA: Ten Speed Press, 2001.

This book includes advice on writing for skim-reading and writing for the world along with advice on writing keywords, alternative text for images, and titles.

McGovern, Gerry, and Rob Norton. *Content Critical: Gaining Competitive Advantage through High-Quality Web Content*. London: Pearson Education Limited, 2002.

While treating the reader and content as crucial to a successful Web site, this book emphasizes editing to control the quality, and navigation and layout to make the content easy to find.

Nielsen, Jakob. *Designing Web Usability: The Practice of Simplicity*. Indianapolis, IN: New Riders Publishing, 1999.

In addition to a major focus on providing quality content on the Web and helping users find what they need, Nielsen devotes a chapter each to accessibility and to international use of the Web.

Price, Jonathan, and Lisa Price. *Hot Text: Web Writing That Works*. Indianapolis, IN: New Riders Press, 2002.

This book offers advice for writing for the Web such as "make text scannable" and "reduce cognitive burdens."

Rubin, Jeffrey. *Handbook of Usability Testing: How to Plan, Design, and Conduct Effective Tests*. New York: John Wiley & Sons, Inc., 1997.

This book is aimed at helping people who have not been trained as human factors engineers and yet need to improve the usability of what they work on. This book provides practical guidance from developing a test plan through conducting the tests to analyzing the results. It includes information about user-centered design.

Glossary

A

abbreviation. A shortened form of a word or phrase, used in place of the whole. Unlike acronyms, abbreviations can end in a period and might not be all uppercase. For example, Calif. is an abbreviation for California.

abstract. *(noun)* A brief summary of the main points that are in a document or presentation. *(adjective)* A characteristic of information in which concept, task, or reference topics are presented as theoretical and unrelated to real experience.

accessibility. In information technology, accommodations to enable people with disabilities (such as those that affect vision, hearing, mobility, and attention) to use computer software and hardware.

accuracy. Freedom from mistake or error; adherence to fact or truth.

acronym. A word formed from the first letter of each part or of each major part of a compound term. For example, CAD stands for computer-aided design.

active style. A manner of expression that uses active voice and addresses the user directly.

active voice. A grammatical form in which the doer of the action is expressed as the subject rather than not at all or as an agent. Contrast with *passive voice*.

ambiguous. Capable of being understood in two or more senses.

analogy. An explicit comparison based on a resemblance between things that are otherwise not alike. See also *metaphor, simile.*

animation. The motion that is simulated by displaying a series of images. Animation can be a major part of a multimedia presentation.

antecedent. A word, phrase, or clause that a pronoun refers to.

artificial task. A task that is imposed by a product or presented in terms of using a product component. For example, "using the CNTREC utility" is artificial, but "counting records with the CNTREC utility" is not. Contrast with *real task*.

audience. A group of readers who have similar tasks and a similar background; they might have differing expertise in doing these tasks.

audience analysis. An investigation of user characteristics such as education and training, level of experience with the subject or with related subjects, and work environment.

B

background. The area behind the text or underlying a graphic.

balance. *(noun)* A pleasing arrangement of visual elements. *(verb)* To integrate visual elements in an effective way. See also *visual element*.

beta test. *(noun)* A test of a pre-release version of a product, usually involving users outside the company. An earlier test, usually involving users within the company, is an alpha test. In an instrumented beta test, hooks within the code enable developers to monitor the use of the product. *(verb)* To conduct a beta test.

blank space. The space that is empty on a screen or page. It can be small, such as the amount of an indention, or large, such as the whole space not filled by type or visual elements. Distinguish from *blank character* or *space character,* which is the space that you get when you press the space bar on a keyboard. See also *white space*.

boilerplate. Any standardized text (such as for legal notices) that is provided to ensure consistency across documents and products.

branch. *(noun)* A jump to another place in the information, usually marked by a link or cross-reference. A user can choose to go to the other information or not to go. *(verb)* To follow a link or cross-reference, leaving a topic and going to a related topic. See also *link*.

breadcrumb trail. A set of links that show the Web pages that have been visited to reach the current page. Such links facilitate returning to an earlier page.

bulleted list. See *unordered list*.

C

capture. See *screen capture*.

chunk. *(noun)* A logical unit of information that addresses only one subject and allows modularity of the information. *(verb)* To group information into discrete parts based on content, and to label each part, or chunk.

clarity. Freedom from ambiguity or obscurity; the presentation of information in a way that helps users understand it the first time.

code example. One or more lines of code, in any programming language, that users can use as is or can adapt to their own situation.

cohesive. Characterized by a readily understandable flow of information from sentence to sentence, from paragraph to paragraph, and from text to a graphic or table and back.

competitive evaluation. In the context of technical information, an analysis of the information of a competitive product or of information that is considered similar, to determine relative strengths and weaknesses.

completeness. The inclusion of all necessary parts—and only those parts.

conceptual information. Information that pertains to design, ideas, relationships, and definitions. See also *task information, reference information.*

concreteness. The inclusion of appropriate examples, scenarios, similes, analogies, specific words, and graphics.

conditional text. Designated text that is included or excluded based on conditions that the writer specifies. This type of text helps enable the reuse of source in multiple contexts.

container. In a topic hierarchy, a topic that gathers together related subtopics. A container can have substantive information itself (often an overview of the subtopics) or can serve only as a necessary heading in the hierarchy.

content pane. Information that is provided in a separate pane from the links to that information. See also *navigation pane.*

contextual information. The online information that is relevant to where the user is in a software program. Contextual information typically includes information about the window or field for which the user requested help. Sometimes called *context-sensitive help.*

control. See *interface control.*

cross-reference. A citation at one place (as in a document) to another place that contains relevant information. See also *link.*

cue card. A type of online information that provides an overview of possible tasks and guides users to instructions for the task that they choose.

D

direct word. A word that is straightforward and easily understood. Often, such a word is short and derives from Anglo-Saxon rather than from Latin.

E

element. A unit that physically constitutes technical information. Such a unit can be, for example, a word, heading, paragraph, list, or tutorial.

entry point. An element that marks the presence and location of information. An entry point can be, for example, a heading, index, hypertext link, or table of contents.

evaluation test. A type of usability test that is done early in a development cycle to determine whether a design will probably meet usability criteria if fully implemented. Contrast with *validation test.*

example. A representative of a set of things. In technical writing, an example can be code, text (such as a scenario), or a graphic.

experience level. A characteristic of users that deals with their first-hand knowledge of the subject as opposed to their education and training.

expert user. A person whose experience and training makes the person comfortable using a tool and able to handle most tasks. Compare with *novice user, user*.

F

field test. *(noun)* A test of a product in an actual or simulated work environment before the product is released. *(verb)* To conduct such a test.

figure. An illustration or graphic that is inserted into a block of textual information to complement or clarify the information. Figures are usually, but not always, labeled. See also *graphic, illustration*.

follow-on task. A task that users need to do after they do a different task. Sometimes called *postrequisite*. Contrast with *prerequisite*.

font. A complete set of letters, numbers, and punctuation of a particular style (such as italics for a *typeface*) and size; one size of a typeface. See also *typeface*.

foot. See *running foot*.

G

gender-neutral language. Words (such as the pronoun *they* and the noun *person*) that do not specify a feminine or masculine gender.

gerund. A word that ends in *-ing* and functions as a noun. A gerund can have objects, complements, or modifiers.

globalization. The process of creating a product that can be used in many cultural contexts without modification. Such a product can process multilingual data and present culturally correct information (such as collation, date format, and number format). See also *internationalization, localization*.

graphic. A self-contained nontextual visual element that is copied (such as a screen capture) and conveys an image. See also *figure, illustration*.

graphical element. Any nontextual visual element that is intentionally added to text to enhance its visual impact. This term applies to graphics, illustrations, or visual organizers such as rules to separate running heads from text and bullets to present parallel items. See also *visual element*.

graphical user interface (GUI). A type of user interface that takes advantage of high-resolution graphics. A typical graphical user interface combines graphics, object-action paradigm, pointing device, menu bars, menus, overlapping windows, and icons. See also *user interface*.

guideline. A brief directive about what to do or not do (in this book, particularly to achieve the characteristics of quality).

H

head. See *running head, heading*.

heading. A prominent phrase that describes the information after it. In this book, phrases that are located in the margin and help retrievability are also considered headings. Contrast with *label*.

help. Discrete pieces of online information about parts of a product such as a window, field, message, or task. Also, the set of discrete pieces of online information as a whole. See also *contextual information*.

hypertext. An organization of online information that is a network where users can browse information by using links. See also *link*.

I

icon. A small graphic symbol on a computer screen that indicates a task that users can do or something (such as a folder or file) that users can use.

illustration. A created piece of artwork (as opposed to a copied piece of artwork such as a screen capture) such as a drawing, chart, diagram, or photograph used to explain, depict, or complement text. See also *figure*, *graphic*.

image map. A graphic that contains more than one hot spot, each of which goes to a different Web page. See also *navigation bar*.

imperative mood. A form of a verb that indicates a command, such as "Press Enter."

imprecise verb. A verb phrase that consists of a weak verb (such as *make* or *have*) and an object that together carry the meaning.

indention. The blank space between a margin and the beginning of an indented line. Also, the act of moving something in from the margin a certain amount.

information architecture. The structure of a Web site or of an information unit as determined by the organization and navigation system. Through metadata and indexing, information architecture can support browsing and searching. See also *metadata*.

information center. A set of online topics presented with a navigation pane on the left and a content pane on the right. Often the information center includes a search engine, and topics are structured as concepts, tasks, and reference.

information type. A category of information that is based on its content and use. The major types that are discussed in this book are task, concept, reference, and contextual. Other types include marketing, tutorial, and wizard. See also *topic*.

information unit. A collection of information as presented to users. For example, an information unit might be a document such as a guide or reference manual, or the set of help that accompanies a product.

interface. The means by which users interact with computer hardware and software.

interface control. In an online interface, a device such as a button or menu that a person uses to do tasks with the software.

internationalization. The design and development of information or of an application so that it can easily be modified to support different national languages and cultural conventions. See also *globalization*, *localization*.

intrapage link. A link that goes to information within the same Web page, rather than to information on another Web page.

J

jargon. Technical terminology or a characteristic idiom of a science, trade, profession, or similar group.

jump. See *branch, link.*

L

label. A heading (usually one word such as Tip, Restriction, or Important) that governs a paragraph within a larger heading structure. Labels do not show up in a table of contents. See also *heading.*

launchpad. In online information or a software interface, an introductory page with explanations and links so that users can get oriented to a product and start doing major tasks in the product.

link. *(noun)* In online information, an opportunity to jump to related information. Links provide the connections among pieces of information in hypertext. *(verb)* To follow a link, leaving a topic and going to another topic. See also *branch, hypertext.*

list of contents. See *table of contents.*

localization. The process of translating text and creating culturally specific settings for a software product. The process can include packaging the product into a national language version. See also *globalization, internationalization.*

M

machine translation. The use of computer software to translate text from one human language to another. Because of its speed and relatively low cost, machine translation is used particularly for translating Web pages. See also *translation memory.*

message. Information that a user does not request but that a product or application displays. Information messages give status. Error messages give a warning or information about an unexpected event.

metadata. A comprehensive definition or description of a set of data. Metadata might provide, for example, keywords about a topic, the date when the topic was first published, and the date when it was last updated.

metaphor. A word, phrase, or visual representation that denotes or depicts one object or idea but suggests a likeness to or analogy with another object or idea. See also *analogy, simile, visual metaphor.*

minimalism. A set of principles for presenting instructional material that supports exploration, uses real tasks, supports recovering from errors, and provides concise information.

mood. A verb form that expresses whether the action is meant to express a fact or another manner such as a command, possibility, or wish. See also *imperative mood.*

multimedia. A computer application that uses any combination of text, image, animation, video, and audio effects to present information to the user.

N

navigate. In online information, to use controls that are available in the interface to display a different portion of the information.

navigation bar. A set of buttons or graphic images that are typically arranged in a row or column and are used to link to major sections (such as Home, Support, Contact Us) on a Web site. See also *image map, navigation pane*.

navigation pane. Links that are provided in a separate pane (often on the left-hand side of a browser window) and serve as a table of contents to topics that are viewed in a content pane. See also *content pane, navigation bar.*

needless repetition. Redundancy in a phrase where two or more words are synonymous or almost synonymous, as in "sequential steps."

negative expression. A phrase that includes negative words such as *no, not, none, never,* or *nothing*.

nonrestrictive clause. A descriptive clause that adds nonessential information about the noun that it modifies. Contrast with *restrictive clause.*

noun string. A set of two or more nouns that together identify one thing.

novice user. Someone who has little or no experience in a particular area. Compare with *expert user* and *user.*

numbered list. See *ordered list.*

O

online information. Information that is available on a computer, usually to accompany a software product. Such information includes, for example, contextual help, information centers, and linear documents that can be viewed online.

ordered list. A set of items that are in sequential (numerical or alphabetical) order. Nested ordered lists usually include an alphabetical format within a numerical format, as appropriate for an outline. Sometimes called *numbered list*. Contrast with *unordered list.*

organization. A coherent arrangement of parts that makes sense to the user.

overly complete. Characterized by too much information, much more than users need. Often the extra information describes product internals or gives too many details for the particular users.

P

pane. A segment of a window that can be scrolled separately and that might have its own title. Selection of items in a pane can cause items in another pane to change.

parallelism. Using the same grammatical structure for similar elements such as items in a list, nouns or adjectives in a series, and compound clauses.

passive voice. A grammatical form in which the doer of the action is expressed as an agent or not at all. Contrast with *active voice.*

pattern. A form or model for dealing with recurring kinds of elements. A pattern might include headings for various types of information and the layout for information under those headings. See also *template*.

PDF. See *Portable Document Format (PDF).*

pixel. An abbreviation of *picture element*: the smallest single point on a display screen that can be assigned an individual color value.

point. A standard unit of type size. The British and American standard is 72 points to an inch.

pop-up help. In online information, brief contextual help that explains interface controls, particular fields, and the valid values. Sometimes called *bubble help*. See also *interface control.*

Portable Document Format (PDF). A file format, developed by Adobe Systems, that captures formatting information from various source formats, enables sending and printing formatted documents, and presents documentation by page.

postrequisite. See *follow-on task.*

prerequisite. *(noun)* A requirement that is needed before a user can do a task. *(adjective)* Of or relating to a task that must be done (or to an object that must be available) before a user can do a different task. See also *follow-on task.*

primary audience. The group of people who (because of their occupation, for example) are expected to use the product most. See also *secondary audience.*

procedure. A set of steps, in a definite order (though some steps can be conditional or optional), for a user to do a task that is well defined, such as creating index entries or deleting a document. See also *process, task.*

process. A series of events, stages, or phases that lead to a particular output or result. The parts of a process must usually be in a definite order, involve more than one person (perhaps also a computer or other machine), and occur over a period that is longer than for a procedure. Some processes are carried out only by a machine or computer. See also *procedure.*

prototype. (*noun*) A mock-up that helps developers experiment with a design for a product, a system, or an interface. A low-tech (or low-fidelity) prototype is made quickly out of temporary materials such as paper. (*verb*) To create a prototype.

pseudo-task heading. A heading that masks a real task or that makes a conceptual or reference topic seem to be a task topic.

Q

quality. Excellence or superiority in kind.

quality characteristic. An attribute that is essential to describing quality.

R

real task. A task that users want to perform, regardless of what product they use to do it. For example, "balancing your checkbook" is a real task; "using InfoBanker's Balance feature" is not. Contrast with *artificial task.*

reference information. A collection of facts (such as terms, statements, commands, rules, messages, names of windows and buttons, and conventions) that is organized for quick retrieval. See also *conceptual information, task information.*

relative clause. A clause that is introduced by a relative pronoun such as *that* or *which.*

resolution. The dimensions in pixels of the display area of a computer monitor. *Low* resolution is 640 pixels wide by 480 pixels high. *High* resolution is 1024 pixels by 768 pixels or higher. Text on a monitor that is set to 640x480 resolution looks larger than the same text looks on the same monitor when it is set to 1024x768. This difference occurs because the pixels that make up the text occupy a smaller relative portion of the screen in the higher resolution format.

restrictive clause. A descriptive clause that is essential to the meaning of the noun that it modifies. Contrast with *nonrestrictive clause.*

retrievability. The presentation of information in a way that enables users to quickly and easily find specific items.

reuse. *(noun)* The repeated use of information, with few or no changes. *(verb)* To use information again with few or no changes, perhaps for a different document or product. See also *single source.*

roundabout expression. A phrase that uses more words than necessary, such as "in the course of" instead of "during."

rule. A line of any thickness (called *weight*) that serves as a nontextual visual element. Such a line can, for example, set off rows in a table, surround a figure, or separate text from a running head.

running foot. A heading, usually in a font that is smaller than the body text font, in the bottom margin of a page of a print document. Running feet (such as a chapter title or number) help users find information.

running head. A heading, usually in a font that is smaller than the body text font, in the top margin of a page of a print document. Running heads (such as a chapter title or number) help users find information.

S

sample. A set of code or data that is provided with a software product to help users try out the product or parts of it. See also *example.*

sans serif. *(noun)* A letter or typeface that has no serifs. *(adjective)* Of or relating to a typeface whose letters do not include small terminal strokes (serifs) at the end of the main strokes of the letters. See also *serif, slab serif.*

scenario. A series of events over time, usually involving a fictitious but realistic set of circumstances.

screen capture. A graphic likeness of a screen or a portion of the screen that is saved in a file for inclusion in technical information. Sometimes called *screen shot.*

screen reader. An assistive device that provides speech output (and sometimes Braille output) while a vision-impaired person uses a software program.

search engine. Software that acts on user input to find likely matches for what the user is seeking. The space being searched can vary from a page or document to a large set of items or even the whole Web. Each search engine uses its own algorithm to create and return the results of a query.

secondary audience. The group of people who (because of their job, for example) are expected to use the product a significant amount of time but less than the primary audience does. Sometimes called *occasional audience.* See also *primary audience.*

semantic. Of or relating to meaning; used here especially as expressed through the elements of technical information. Compare with *syntactic.*

serif. *(noun)* The small terminal stroke at the end of the main stroke of a letter. *(adjective)* Of or relating to a typeface in which all of the letters include small terminal strokes. See also *sans serif, slab serif.*

sidebar. Additional information, often displayed in a box in the margin, that provides details, an example, or a different viewpoint than the primary information.

simile. An explicit comparison (often introduced by *like* or *as*) of two things to show a similarity that they share. See also *analogy, metaphor.*

single source. A firsthand piece of work (often a file) from which more than one type of information is produced, usually for different media. For example, a writer can use the same task file in both a help system and a printed document. See also *reuse.*

site map. A hierarchical rendering of the pages of a Web site as links, to help users find information. See also *table of contents.*

slab serif. *(noun)* A heavy or squared serif that is almost the same thickness as the main stroke to which it is attached. *(adjective)* Of or relating to a typeface with thick, squared serifs. See *sans serif, serif.*

source. (1) The start for a link in online information or for a cross-reference in print information. (2) A file or its contents that can be transformed for presentation in various forms. See also *single source, target.*

style. (1) Correctness and appropriateness of writing conventions and of words and phrases. (2) A convention with respect to spelling, punctuation, capitalization, or typographic arrangement and display.

style guide. A collection of rules and guidelines to establish what usage is correct and incorrect, good and bad, or just preferred for the sake of consistency. Usually a style guide contains many style guidelines and examples. Sometimes called *style manual*. See also *style guideline.*

style guideline. A brief directive about what to do in a specific situation affecting the appearance or sense of information.

style sheet. (1) A definition of the appearance of a document in terms of typefaces, fonts, spacing, and other visual elements. (2) A quick reference document that summarizes or highlights decisions about the stylistic conventions for a particular project. See also *style guide.*

subject. (1) The part of a clause on which the rest of the clause is predicated. (2) The area of concern in a topic or document. See also *topic.*

subtask. A portion of a long task that has been broken down into discrete subsections. A subtask has enough information (such as rationale and steps) that it makes sense as a discrete topic. See also *task.*

syntactic. Of or relating to syntax as the harmonious arrangement of elements. Compare with *semantic*.

syntactic cue. A grammatical element of language (such as word order, verb endings, articles, and conjunctions) that helps users correctly identify parts of speech and analyze the structure of a sentence.

T

table of contents. A structured list of headings that represent the subjects (usually in a linear document), to help users find information. An online table of contents has a link for each heading, and a print table of contents has a page number for each. See also *navigation pane, site map.*

target. The destination for a link or cross-reference. See also *source.*

task. An activity, physical or mental, that is done with a product (such as editing a file) or for a product (such as installing and maintaining the product). See also *procedure, subtask.*

task analysis. The process of determining primarily who does a task and how, when, where, and why.

task help. Online information that provides guidance about how to do a task with a product.

task information. Information that makes clear how to do a task. It includes procedures, and can also include items such as examples and tips. See also *conceptual information, reference information.*

task orientation. A focus on helping users complete tasks that are associated with a subject or product in relation to their jobs.

technical information. The information on a technical subject, usually for a particular audience and for a stated purpose. Such information can accompany a product or describe something in a science (whether computer science, physical science, or social science), trade, or profession.

technical writing. The process of writing, reviewing, and revising technical information.

template. Something, such as a file with sample headings, that serves as a pattern for creating a similar thing. See also *pattern.*

term. A word or expression that is specific to a science, art, profession, or subject.

tone. A manner of expression in writing that conveys the viewpoint of the writer. In technical writing, a tone is more likely to be serious and authoritative than humorous and condescending, for example.

topic. A chunk of information about a particular subject. Sometimes called *article, module, chunk.* See also *subject.*

translation memory. The software that stores matching segments of a source language and a target language as translated by a human translator for reuse. See also *machine translation.*

typeface. A complete set of letters, numbers, and punctuation of a particular design, such as Helvetica or Times Roman. See also *font.*

type size. The nominal size of a font, in points. This size usually includes a small amount of blank space, or *shoulder,* above and below the actual printed character. Thus, the printed words in a line of 12-point type might be only 10.5 points high. See also *font, point.*

U

UI control. See *interface control.*

unified example. A set of examples that centers around certain data. Pieces of a unified example can be spread throughout a document.

unordered list. A set of items that are arranged in an order that does not require a sequential (numerical or alphabetical) format. Sometimes called *bulleted list.* Contrast with *ordered list.*

usability. The ease with which people in a defined group can learn and use a product (including the information) to accomplish certain tasks.

usability test. A test of the interaction between the user and the product or information for the product.

usability walkthrough. A usability inspection approach in which evaluators (perhaps users) do a set of tasks to determine the ease of learning and using a product interface.

user. A person who comes into contact with a product through doing any of the following tasks with or for the product: buying, evaluating, planning, learning, installing, migrating, using, operating, administering, developing applications, customizing, diagnosing, or maintaining. See also *expert user, novice user.*

user interface. The area where a user and an object come together to interact; as applied to computers, the ensemble of hardware and software that enables a user to interact with a computer program. See also *graphical user interface (GUI).*

vague referent. The noun, phrase, or clause that a pronoun refers to but is hard to determine. See also *antecedent*.

validation test. A type of usability test that is done late in a development cycle to determine whether an implementation meets usability criteria. Contrast with *evaluation test.*

visual effectiveness. Attractiveness and enhanced meaning of information through use of layout, illustrations, color, type, icons, and other graphical devices.

visual element. Any component that affects the visual impact of the information. A visual element could be, for example, highlighting, a heading, table, or illustration. See also *graphical element*.

visual metaphor. An implied likeness between an image or images and familiar objects or concepts in a particular domain. Visual metaphors are not necessarily intuitive. See also *metaphor.*

walkthrough. See *usability walkthrough.*

Web page. A document on the Web that is identified by a unique URL (uniform resource locator). See also *Web site.*

Web site. A collection of directories and files on a server that can be accessed by users with Web browsers. See also *Web page.*

white space. The portion of a printed page or online window that is left blank. See also *blank space.*

window. A scrollable viewing area (usually rectangular) on a display screen.

wizard. A type of interface that leads users through well-defined steps to produce a result, such as creating a chart from information in a database.

Index

A

B

C

E

F

J

L

M

N

O

S

T

U

V

W

About the authors

The authors have served on the Editing Council at IBM Silicon Valley Laboratory in San Jose, California, an organization dedicated to excellence in technical information.

Gretchen Hargis is a technical manager at IBM for a group that provides user assistance and user-centered design for application development tools. She was a technical editor and writer and a pioneer of IBM Darwin Information Typing Architecture (DITA).

Michelle Carey is a technical writer at IBM and a technical writing instructor at University of California Santa Cruz Extension. She is an expert on topic-based information systems and on writing for international audiences.

Ann Kilty Hernandez is a technical editor at IBM and has been a technical writer, manager, and marketing specialist. She was a co-author of *An Introduction to DB2 for OS/390* and contributed to its next edition, *The Official Guide to DB2 UDB for z/OS*.

Polly Hughes, now retired from IBM, worked as a visual designer for technical information and software interfaces and as a technical writer.

Deirdre Longo is a technical editor and writer at IBM who edits product interfaces and writes customer information, mostly for content management products.

Shannon Rouiller is a technical editor at IBM who has written and edited topic-based information systems, books, contextual help, wizards, and interfaces for products that are marketed worldwide. She co-authored *Designing Effective Wizards*.

Elizabeth Wilde is a technical editor at IBM and a leader in developing quality metrics and quality assurance processes for technical documentation. She also educates writers and editors throughout IBM on developing user-centered information.